ELEMENTARY AND SECONDARY EDUCATION FOR SCIENCE AND ENGINEERING

A Technical Memorandum

Daryl E. Chubin
Project Director

HEMISPHERE PUBLISHING CORPORATION
A member of the Taylor & Francis Group
New York Washington Philadelphia London

ELEMENTARY AND SECONDARY EDUCATION FOR SCIENCE AND ENGINEERING:
A Technical Memorandum

B+T App 4-3-90

1 2 3 4 5 6 7 8 9 0 BRBR 8 7 6 5 4 3 2 1 0 9

Cover design by Debra Eubanks Riffe.
A CIP catalogue record for this book is available from the British Library.

Library of Congress Cataloging-in-Publication Data

Chubin, Daryl E.
Elementary and secondary education for science and engineering / OTA project staff ; John Andelin, Assistant Director, OTA, Science, Education, and Natural Resources Division ; Nancy Carson, Science, Education, and Tansportation, Program Manager.
p. cm.
Written by Daryl E. Chubin, Richard Davies, and Lisa Heinz.
Includes bibliographical references.

1. Science—Study and teaching. 2. Mathematics—Study and teaching. I. Davies, Richard, date. II. Heinz, Liza C., date. III. United States. Congress. Office of Technology Assessment. IV. Title.
LB1585.C44 1990
507.1'2—dc20 88-600594
ISBN 1-56032-045-1 CIP

Contents

Foreword

Choice, chance, opportunity, and environment are all factors that determine whether or not a child will grow up to be a scientist or engineer. Though comprising only 4 percent of our work force, scientists and engineers are critical to our Nation's continued strength and vitality. As a Nation, we are concerned about maintaining an adequate supply of people with the ability to enter these fields, and the desire to do so.

In response to a request from the House Committee on Science and Technology, this technical memorandum analyzes recruitment into and retention in the science and engineering pipeline. *Elementary and Secondary Education for Science and Engineering* supplements and extends OTA's June 1988 report, *Educating Scientists and Engineers: Grade School to Grad School.*

Students make many choices over a long period, and choose a career through a complicated process. This process includes formal instruction in mathematics and science, and the opportunity for informal education in museums, science centers, and recreational programs. The influence of family, teachers, peers, and the electronic media can make an enormous difference. This memorandum analyzes these influences. Because education is "all one system," policymakers interested in nurturing scientists and engineers must address the educational environment as a totality; changing only one part of the system will not yield the desired result.

The Federal Government plays a key role in sustaining educational excellence in elementary and secondary education, providing effective research, and encouraging change. This memorandum identifies pressure points in the system and strengthens the analytical basis for policy.

JOHN H. GIBBONS
Director

Elementary and Secondary Education for Science and Engineering Advisory Panel

Neal Lane, *Chairman*
Provost, Rice University, Houston, TX

Amy Buhrig
Specialist Engineer
Artificial Intelligence
Boeing Aerospace Corp.
Seattle, WA

David Goodman
Deputy Director
New Jersey Commission on Science and Technology
Trenton, NJ

Irma Jarcho
Chairman
Science Department
The New Lincoln School
New York, NY

Hugh Loweth
Consultant
Annandale, VA

James Powell[1]
President
Franklin and Marshall College
Lancaster, PA

Rustum Roy
Evan Pugh Professor of the Solid State
Materials Research Laboratory
Pennsylvania State University
University Park, PA

Bernard Sagik
Vice President for Academic Affairs[2]
Drexel University
Philadelphia, PA

William Snyder
Dean, College of Engineering
University of Tennessee
Knoxville, TN

Peter Syverson
Director of Information Services
Council of Graduate Schools in the United States
Washington, DC

Elizabeth Tidball
Professor of Physiology
School of Medicine
The George Washington University
Washington, DC

Melvin Webb
Biology Department
Clark/Atlanta University
Atlanta, GA

F. Karl Willenbrock
Executive Director
American Society for Engineering Education
Washington, DC

Hilliard Williams
Director of Central Research
Monsanto Company
St. Louis, MO

Dorothy Zinberg
Center for Science and International Affairs
Harvard University
Cambridge, MA

[1]Currently President, Reed College.

[2]Currently Professor of Bioscience and Biotechnology.

NOTE: OTA appreciates and is grateful for the valuable assistance and thoughtful critiques provided by the advisory panel members. The panel does not, however, necessarily approve, disapprove, or endorse this technical memorandum. OTA assumes full responsibility for the technical report and the accuracy of its contents.

Elementary and Secondary Education for Science and Engineering OTA Project Staff

John Andelin, *Assistant Director, OTA*
Science, Information, and Natural Resources Division

Nancy Carson
Science, Education, and Transportation Program Manager

Daryl E. Chubin, *Project Director*

Richard Davies, *Analyst*

Lisa Heinz, *Analyst*

Marsha Fenn, *Technical Editor*

Madeline Gross, *Secretary*

Robert Garfinkle, *Research Analyst*

Other Contributors

The following individuals participated in workshops and briefings, and as reviewers of materials produced for this technical memorandum. OTA thanks them for their contributions.

Patricia Alexander
U.S. Department of Education

Rolf Blank
Council of Chief State School Officers

Joanne Capper
Center for Research Into Practice

Dennis Carroll
U.S. Department of Education

Ruth Cossey
EQUALS Program
University of California, Berkeley

William K. Cummings
Harvard University

Linda DeTure
National Association of Research in Science Teaching

Marion Epstein
Educational Testing Service

Alan Fechter
National Research Council

Michael Feuer
Office of Technology Assessment

Kathleen Fulton
Office of Technology Assessment

James Gallagher
Michigan State University

Samuel Gibbon
Bank Street College of Education

Dorothy Gilford
National Research Council

Kenneth C. Green
University of California, Los Angeles

Michael Haney
Montgomery Blair Magnet School

Thomas Hilton
Educational Testing Service

Lisa Hudson
The Rand Corp.

Paul DeHart Hurd
Stanford University

Ann Kahn
National Parent Teacher Association

Daphne Kaplan
U.S. Department of Education

Susan Coady Kemnitzer
Task Force on Women, Minorities, and the Handicapped in Science and Technology

Dan Kunz
Junior Engineering Technical Society

Cheryl Mason
San Diego State University

Barbara Scott Nelson
The Ford Foundation

Gail Nuckols
Arlington County School Board

Louise Raphael
National Science Foundation

Mary Budd Rowe
University of Florida

James Rutherford
American Association for the Advancement of Science

Vernon Savage
Towson State University

Anne Scanley
National Academy of Sciences

Allen Schmieder
U.S. Department of Education

Susan Snyder
National Science Foundation

Julian Stanley
The Johns Hopkins University

Harriet Tyson-Bernstein
Consultant

Bonnie VanDorn
Association of Science-Technology Centers

Betty Vetter
Commission on Professionals in Science and Technology

Leonard Waks
Pennsylvania State University

Iris R. Weiss
Horizon Research, Inc.

John W. Wiersma
Huston-Tillotson College

Reviewers

Richard Berry
Consultant

Audrey Champagne
American Association for the Advancement of Science

Edward Glassman
U.S. Department of Education

Shirley Malcom
American Association for the Advancement of Science

Willie Pearson, Jr.
Office of Technology Assessment

Linda Roberts
Office of Technology Assessment

George Tressel
National Science Foundation

Preface

This technical memorandum augments OTA's report, *Educating Scientists and Engineers: Grade School to Grad School*,[1] focusing on the factors that prompt students to plan science and engineering careers during elementary and secondary education, and the early stages of higher education. While examining the problems and opportunities for students, OTA offers no comprehensive assessment of the system of American public education. Rather, it takes this system as the context for understanding, and proposes changes in preprofessional education.

Most educators and parents regard science and mathematics as basic skills for *every* high school graduate. By upgrading mathematics and science literacy—making more graduates proficient in these subjects—most believe that the pool of potential scientists and engineers would be larger and more diverse. At the same time, broader application of basic skills in mathematics and science would benefit the entire U.S. work force. Perhaps then concern for the future supply of scientists and engineers, as one professional category of workers among many, would recede as an urgent national issue.

As this is not the case, the problem of educating scientists and engineers is unabating; the scrutiny of schools, teaching standards, and student outcomes is intensifying; and calls for improved Federal action grow louder. As Paul Gray, President of the Massachusetts Institute of Technology said: "Americans must come to understand that engineering and science are not esoteric quests by an elite few, but are, instead, humanistic adventures inspired by native human curiosity about the world and desire to make it better."[2]

OTA takes a long view of the science and engineering pipeline and does not dwell solely on those whose scientific talents are manifested at early ages. The science and engineering pipeline includes all students during elementary and much of secondary schooling. But as students move toward undergraduate and graduate school, smaller proportions form the talent pool. Students make choices over long periods and are influenced by many factors; this complicates analysis and makes it difficult to ascertain specific influences or their degree of impact on careers.

Although most students' career intentions are ill-formed, some decide to pursue science and engineering early in life and stick with that decision. This "hard-core group" is joined by many companions later on. **Chapter 1, Shaping the Science and Engineering Talent Pool**, concerns both the hard core and those whose plans are more malleable. As this latter group is uncertain about what major to choose, it may be more susceptible to parents' wishes, financial incentives, and the attractiveness of science and engineering careers. Whether students respond to a professional "calling" or hear the call of the marketplace, they are lured to some careers and away from others—and schools are agents of this allure.

For many children, the content of mathematics and science classes and the way these subjects are taught critically affect their interest and later participation in science and engineering. **Chapter 2, Formal Mathematics and Science Education**, reviews concern over the pace and sequence of the American mathematics and science curriculum, the alleged dullness of many science textbooks, and the extent to which greater use of educational technology, such as computers, could improve the teaching of mathematics and science.

This responsibility falls primarily on the teaching profession, together with school districts and teacher education institutions. **Chapter 3, Teachers and Teaching**, discusses predicted shortages of mathematics and science teachers, and concern about the poor quality of teacher training and inservice programs in all subjects. The quality of teaching, in the long run, depends on the effectiveness of teachers, the adequacy of their numbers, and the extent to which they are supported by principals, curriculum specialists, technology and materials, and the wider community. Teachers of mathematics and science need to be educated to high professional standards and, like

[1]U.S. Congress, Office of Technology Assessment, *Educating Scientists and Engineers: Grade School to Grad School*, OTA-SET-377 (Washington, DC: June 1988).

[2]Paul E. Gray, "America's Ignorance of Science and Technology Poses a Threat to the Democratic Process Itself," *The Chronicle of Higher Education*, May 18, 1988, p. B-2.

members of other professions, they also need to update their skills periodically.

At the same time, research on teaching of mathematics and science suggests that some techniques, not widely used in American schools, can improve achievement, transmit more realistic pictures of the enterprise of science and mathematics, and broaden participation in science and engineering by women and minorities. **Chapter 4, Thinking About Science Learning**, asks: How can more students be successful in science and mathematics? Does science and mathematics education search for and select a particular type of student, one with a certain learning style? This chapter describes other efforts to correct misconceptions (held both by students and teachers), spur creativity, develop "higher order thinking skills," and to place more students on pathways to learning science and mathematics.

The out-of-school environment offers opportunities to raise students' interest in and awareness of science and mathematics. **Chapter 5, Learning Outside of School**, highlights "informal education" activities that draw strength from the local community—churches, businesses, voluntary organizations, and their leaders. All are potential agents of change. All are potential filters of the images of science and scientists—often negative, almost always intimidating—transmitted by television and other media. Science centers and museums, for example, can awaken or reinforce interest, without raising the spectre of failure for those who lack confidence in their abilities. Intervention programs, aimed especially at enriching the mathematics and science preparation of women, Blacks, Hispanics, and other minorities, can rebuild confidence and interest, tapping pools of talent that are now underdeveloped.

The problems that face mathematics and science education in the schools are complicated and interrelated. **Chapter 6, Improving School Mathematics and Science Education**, proposes a systemic approach to these problems, requiring a constellation of solutions. Reforms, however, tend to be incremental. Change in any one aspect of mathematics and science teaching, such as coursetaking, tracking, testing, and the use of laboratories and technology, is constrained by other aspects of the system, such as teacher training and remuneration, curriculum decisions, community concerns and opinions, and the influences of higher education.

Finally, this report illuminates the gulf between knowing and doing, between recognizing "what works" and replicating it. On many educational issues, experts are groping to specify the boundaries of their ignorance; on others, there are massive data on causes and effects, but little wisdom on how to implement change. It is this latter need that invites Federal initiative, whether "seeding" a program or showing how various partners might collaborate to approach a nagging problem in a novel way. The Federal Government is pivotal for sustaining the policy climate and catalyzing change: If there is a national will, there *is* a way.

Chapter 1

Shaping the Science and Engineering Talent Pool

Photo credit: William Mills, Montgomery County Public Schools

CONTENTS

Box

Figures

Tables

Chapter 1

Shaping the Science and Engineering Talent Pool

To the Committee [the President's Science Advisory Committee], enhancing our manpower supply is primarily a matter of quality not quantity, not a matter of diverting more college students to science and engineering, but of providing for more students who have chosen this career route the opportunity to continue their studies.

Jerome Wiesner, 1963

All scientists and engineers were once children. Families, communities, and the ideas and images presented by books, magazines, and television helped form their attitudes, encouraged their interest, and guided them to their careers. Schools refined their talents and interests, prepared them academically, and gave them confidence by recognizing their aptitude and achievement.

The importance of families and other out-of-school influences on this process can hardly be overemphasized. Students form opinions and learn about science and scientists from families and friends, from the media, and from places such as science centers and museums, summer camps, and summer research experience. Equally, families, friends, and the media can dull interest in science. Nevertheless, it is largely schools, through preparatory courses in mathematics and science, testing methods, and teaching practices, that determine how many young people will prepare sufficiently well for science and engineering careers (and for other careers). It is in the Nation's interest to see that schools provide the widest possible opportunities, and the best possible educational foundations for the study of science and engineering.[1] Some schools meet these goals, but many do not. A small minority of determined students no doubt can triumph over poor teaching, inadequate course offerings, and overrigid or biased ability grouping or tracking. For most—even some of the most talented—these failings of the schools can kill interest and waste talent.

Of particular concern are women and some racial and ethnic minorities, who together represent a large reservoir of untapped talent. Minorities in particular will make up larger proportions of the population in the future. Identifying and motivating talented minority youngsters is an increasingly important necessity for schools.

Concern about the quality of science and mathematics education is also part of a broader concern about the Nation's schools. The objectives, funding, quality, and content of American education are all currently being debated, and a variety of remedies have been proposed.[2]

[1]Unless otherwise noted, this technical memorandum is concerned exclusively with students' interest in *natural* science and engineering subjects. The adequacy of the preparation of future social scientists is not considered.

[2]National Commission on Excellence in Education, *A Nation At Risk* (Washington, DC: U.S. Government Printing Office, 1983). The sequel, Secretary of Education William Bennett's *American Education: Making It Work*, does not quell the concern. Also see National Science Board, *Educating Americans for the 21st Century* (Washington, DC: Commission on Precollege Education in Mathematics, Science, and Technology, 1983); Paul E. Peterson, "Economic and Policy Trends Affecting Teacher Effectiveness in Mathematics and Science," *Science Teaching: The Report of the 1985 National Forum for School Science*, Audrey B. Champagne and Leslie E. Hornig (eds.) (Washington, DC: American Association for the Advancement of Science, 1986).

PREPARING FOR SCIENCE AND ENGINEERING CAREERS

In theory, the preparation of those intending to become scientists or engineers is assumed to be more intensive than that required of the entire school population. In practice, the interest of both groups must be stimulated. All students need fundamental preparations in mathematics and science in the early years of school. The broad goal of improving the understanding of science and technology by all high school graduates (often called scientific or technological literacy) is very closely tied to that of educating future scientists and engineers. Only at the high school level, where the courses chosen by each stream diverge significantly, does this tie begin to loosen.

The Pipeline Model

The path by which young people approach careers in science and engineering is commonly visualized as a kind of pipeline. Students enter the pipeline as early as third grade, where they begin to be channeled through a prescribed level and then sequence of preparatory mathematics and science courses. This channeling pervades the undergraduate and graduate studies that train and credential them as professionals. Many students drop out along the way, losing interest or falling behind in preparation. Few, it is generally thought, enter the pipeline after junior high school. In fact, students' intentions remain volatile until well past high school, with substantial numbers entering the pipeline (by choosing science and engineering majors) by their sophomore year of college. Many late entrants are relatively ill-prepared, however, and may suffer attrition on their path to a baccalaureate.

The pipeline model projects the supply of future scientists and engineers on the basis of the demographic characteristics of successive birth cohorts. But this process is complicated. Career choices, perceptions of opportunities, knowledge of employment markets, and other influences draw students into and out of the talent pool. Changing educational standards and practices also influence the size of this pool.

The education system thus can be thought of as a kind of semipermeable, or leaky, pipeline, with many points of entry and exit through which different students pass with different degrees of ease. Entrance to and persistence in this semipermeable pipeline vary with job opportunities as well as with individual propensities toward knowledge and personal fulfillment. In fact, fields of science and engineering offer widely different incentives that reflect economic and social trends. Thus, the semipermeable pipeline should be thought of as branched, with openings into diverse job markets and careers.

Influences on the Future Composition of the Talent Pool

These observations suggest that the talent pool *can* be enlarged, and changing demographics suggest that it *must* be enlarged. If schools were more generous in identifying talent, and urged college-preparatory mathematics and science courses on more students (not just those who believe they "need" them for career purposes), both the size and quality of the talent pool would be improved. Our scientists and engineers would be more numerous, better trained, and drawn from a population more representative of American society.

Yet the Federal Government is limited in its impact on elementary and secondary education: schools are State and local responsibilities. Research, curriculum development, demonstration projects, equity, and leadership ("jawboning") are traditional Federal roles, but applying the results to classrooms is up to the State education authorities and the 16,000 local public school boards. Change, in this environment, is slow to come. Another reason for a limited Federal role is that science and mathematics education is but one part of a constellation of educational activities. Teaching, testing, and tracking practices are deeply embedded within the schools. Improvements in science and mathematics education are closely related to reforms in education overall.

Demographic Trends

Almost all of those who will be the college freshmen of 2005 were born by 1987. Knowledge of current birth patterns allows us to make very reliable forecasts of the size and the racial and ethnic composition of the college-age population for the next 18 years, and very good estimates even farther into the future.[3]

There are two prominent trends already apparent. The first is that the number of 18-year-olds is declining, and will bottom out by the mid-1990s. The second is that racial and ethnic minorities today form an increasing proportion of the school age population.[4] However, the absolute number of Black 18-year-olds is currently falling, just like the number of white 18-year-olds (but the Black birthrate remains higher).

In general, America's schoolchildren will look increasingly different from past generations. As Harold Hodgkinson writes:

> . . . there will be a Black and Hispanic (Mexican-American) Baby Boom for many more years. Hispanics will increase their numbers in the population simply because of the very large numbers of young Hispanic females. These population dynamics already can be seen in the public schools. Each of our 24 largest school systems in the U.S. has a "minority majority," while 27 percent of all public school students in the U.S. are minority. . . . Looking ahead, we can project with confidence that by 2010 or so, the U.S. will be a nation in which *one of three* will be Black, Hispanic, or Asian-American.[5]

What is unclear is how this demographic transition will translate into college attendance and pursuit of science and engineering degrees. Variations by region and social class, as well as ethnicity, complicate predictions. These are some current trends:

- A continued drop in the number of minority high school graduates who enter college, due to the increased attractiveness of the Armed Forces and disillusionment with the value of a college degree in today's job market. (Overall, college *attendance* is currently holding level, owing to the increased numbers of older students enrolling and a current small increase in the number of high school graduates.)
- A continuing increase in the size of the Black middle class, whose children enroll in higher education at about the same rate as do the children of white middle-class families.
- Continuing high dropout rates for Hispanics, only about 40 percent of whom complete high school.
- Rising concentrations of Hispanics in the Southwest and California (enrollment in California's public schools is already "minority majority").
- Significant increases in the number of high school graduates in the West and Florida during the next 20 years, along with declines of as much as 10 to 20 percent in New England, the Midwest, and the Mountain States.[6]

Educational Opportunity and the Demographic Transition

The participation of females, Blacks, and Hispanics in science and engineering has increased substantially during the last 30 years, but is still small relative to their numbers in the general population.[7] Success in preparation for science and

[3]The actual size of the college freshman class is also determined by the number of older people that enter higher education. At the moment, many people older than the traditional college-going age are indeed entering higher education. In 1985, over 37 percent of those enrolled in college were 25 years of age and older. U.S. Department of Education, Office of Educational Research and Improvement, Center for Education Statistics, *The Condition of Education: A Statistical Report* (Washington, DC: 1987), p. 122.

[4]U.S. Congress, Office of Technology Assessment, *Educating Scientists and Engineers: Grade School to Grad School*, OTA-SET-377 (Washington, DC: U.S. Government Printing Office, June 1988), pp. 8-9.

[5]Harold L. Hodgkinson, *Higher Education: Diversity Is Our Middle Name* (Washington, DC: National Institute of Independent Colleges and Universities, 1986), p. 9. Asian-Americans are well represented in science and engineering; they are categorically excluded from OTA's discussion of educationally disadvantaged minorities.

[6]Jean Evangelauf, "Sharp Drop, Rise Seen in Graduates of High Schools," *The Chronicle of Higher Education*, May 4, 1988, pp. A28-A43.

[7]U.S. Congress, Office of Technology Assessment, *Demographic Trends and the Scientific and Engineering Work Force—A Technical Memorandum* (Washington, DC: U.S. Government Printing Office, December 1985), ch. 5. Discussion of women and minorities in science and engineering often concerns their low level of participation relative to men and whites. Accurate description of this situation depends on definitions and meanings of the terms "underrepresentation" and "overrepresentation." The benchmark most often cited for an "equitable" level of participation is one where the ethnic, racial, and sex composition of the science and engineering work force closely approximates that of the general population. But there is no analytical reason why such a balance should exist. Still, this social goal encompasses the widely embraced motives of promoting equal opportunity, maximizing utilization of available talent,

(continued on next page)

engineering careers takes commitment, work, and inspiration, all of which the education system is supposed to promote. If achievement testing, tracking, sexism, and racism in the classroom, or some combination of these and other factors, prevent success, it is because the system ignores individual differences in intellectual development and discourages capable students from becoming scientists and engineers. Such an outcome would be tragic for the Nation.

The science and engineering talent pool is not fixed either in elementary or in secondary school. A determined core group is joined by a "swing group" of potential converts to science and by late-bloomers, so that the future supply of students who will take degrees in science and engineering is not determined solely by the size and demographic composition of each birth cohort. The past interest and performance of female and minority students in science and engineering fields is a tenuous basis for concluding that a shortage of scientists and engineers is inevitable. Rather than accept demographic determinism,[8] OTA has chosen to investigate the formation of the science and engineering pool in high school and assess how the structure of schooling identifies, reinforces, and perhaps stifles aspirations to careers in science and engineering.

Photo credit: Lawrence Hall of Science

Schools need to adjust to an increasing proportion of minority children.

(continued from previous page)
and aligning the objectives and conduct of science and engineering with the societal value of broad participation.

Comparisons on minority work force participation should generally be made with regard to age, because racial and ethnic composition varies by birth cohort. Other considerations may include regional demographic variations, enrollment and educational status, and economic status of the reference population. Another difficulty is that "Black," "Hispanic," and "white" are imprecise terms. They are largely an arbitrary, albeit simple, way of classifying a population. There are often bigger differences within each group than there are among the groups. The professional and educational status of various groups deserves a more accurate description than "underrepresentation" and "overrepresentation" convey.

[8]A.K. Finkbeiner, "Demographics or Market Forces?" *Mosaic*, vol. 18, No. 1, spring 1987, pp. 10-17.

HIGH SCHOOL STUDENTS' INTEREST IN SCIENCE AND ENGINEERING

To find out how high school students come to see natural science and engineering as potential careers, OTA analyzed the Department of Education's High School and Beyond (HS&B) database, which describes the progress of a sample of those who were high school sophomores in 1980 by surveying them at 2-year intervals after 1980.[9] Students in the sample were asked each year their planned majors, if they were to attend college.

OTA found that, as high school sophomores in 1980, nearly one-quarter of students were interested in natural science and engineering majors.

[9]Valerie E. Lee, "Identifying Potential Scientists and Engineers: An Analysis of the High School-College Transition," OTA contractor report, 1987. The High School and Beyond database also includes data from a sample of high school seniors in 1980. A followup on both of these cohorts was conducted in 1986, but the results were reported too late to be included in the OTA analysis discussed here. This database also includes information on those planning social science majors, but these have not been considered here. For analysis of the 1972 cohort, see Educational Testing Service, *Pathways to Graduate School: An Empirical Study Based on National Longitudinal Data*, ETS Research Report 87-41 (Princeton, NJ: December 1987). For an inventory of national databases on K-12 mathematics and science education, see app. A.

As seniors, almost as many were still interested in these majors, although their field preferences had shifted somewhat. Two years later, 15 percent of the original group of students were in college and planning science or engineering majors.[10] However, as the following discussion will show, this 15 percent was not simply the remnant of those who had expressed interest earlier. In fact, only about 20 percent of this 15 percent indicated science and engineering majors at all three time points in this survey. In other words, many were either new entrants to the pipeline altogether or students whose interests in both science and nonscience majors were volatile. Another striking finding was the substantial number of "nontraditional" students in that 15 percent; about one-quarter of them had not been in the academic curriculum track of high school, for example.

To explore variations in students' interests in different science and engineering fields, OTA defined four broad field categories—health and life sciences, engineering, computer and information sciences, and physical sciences and mathematics.[11] (See figure 1-1.) The most popular field

[10]Most of the decline in interest in natural science and engineering majors is due to the overall decline in the proportion of the sample going to college. When a more select group is considered—not just high school graduates, but those who are contemplating attending or are in college—the proportion planning science and engineering majors is 27 percent as high school sophomores, 28 percent as seniors, and 24 percent as college sophomores. Unlike the larger sample of all high school graduates, the more select group of college-bound high school graduates decreases in size over time, as some students who contemplate going to college do not attend, or drop out, and are consequently defined out of the sample at those times (in this case, 1980 or 1982). No data are available on the number of students that subsequently graduated in science and engineering. The sample reported here for 2 years after high school graduation will be referred to as "college sophomores," even though some were freshmen or not enrolled continuously in college.

[11]Classification of students among fields was based on questions in the High School and Beyond survey and on a reduction to five field categories: life and health sciences, engineering, computer and information sciences, physical sciences and mathematics, and nonscience majors. Life and health sciences included those intending medical professions, including nursing. "Technology" majors were not included in the engineering category because students indicating this interest tend to pursue vocational training. Social sciences were excluded from the science field categories and included in nonscience majors. Thus, for this analysis, nonscience majors included business, preprofessional, social sciences, vocational/technical, and humanities. Social science majors (including psychology) are considered in U.S. Congress, Office of Technology Assessment, "Higher

(continued on next page)

Figure 1-1.—Popularity of Selected Fields Among 1982 High School Graduates Intending to Major in Natural Science or Engineering, 1980-84[a]

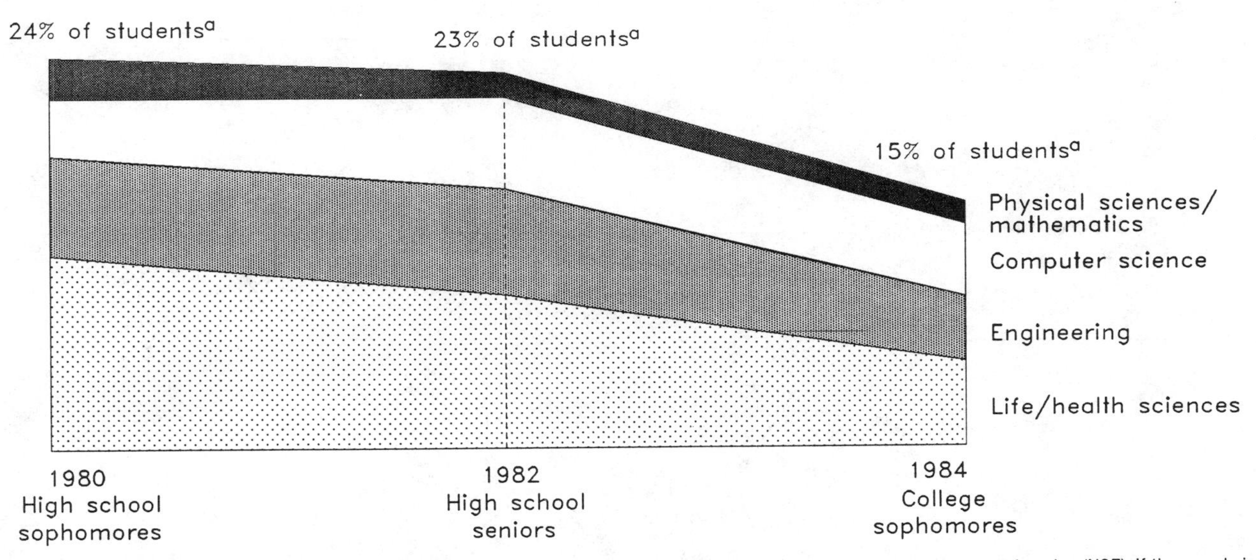

[a]Percent of a nationally representative sample of 1982 high school graduates (n=10,739) who plan to major in natural science or engineering (NSE). If the sample is restricted to an increasingly select group of only those high school graduates who plan to go or have not ruled out going to college, the percent interested in NSE increases: 27% of high school sophomores (n=9,536), 28% of high school seniors (n=8,817), and 24% of college sophomores (n=6,583). (The number of college-bound students decreases with time because some students planning college do not attend or drop out.)

SOURCE: Valerie E. Lee, "Identifying Potential Scientists and Engineers: An Analysis of the High School-College Transition. Report 1: Descriptive Analysis of the High School Class of 1982," OTA contractor report, July 20, 1987, pp. 21-22. Based on data from U.S. Department of Education, High School and Beyond survey.

category at all three time points was health and life sciences. The next most popular field was engineering. In the college sophomore year, engineering was overtaken by computer and information science, which was generally the third most popular field. Physical sciences and mathematics were the least popular potential college majors.

Persistence and Migration in the Pipeline

Although these data confirm the net loss of students from intended science and engineering majors, they need to be supplemented by data on the number of students who moved into and out of the pipeline. Figure 1-2 documents the flow of students out of and into science and engineering fields, and into the four field categories, over the intervals formed by the three survey points (sophomore year of high school, senior year of high school, and sophomore year of college).

Between sophomore and senior years of high school, the figure shows that, in every field category, more students came *into* each of the four fields than persisted in them. Movement in from the nonscience category was more common than movement between science field categories. Patterns of persistence were less clear during the transition from senior year of high school to sophomore year of college. In all field categories except engineering, more students moved in than persisted.

Overall patterns of persistence are presented in figure 1-3, which shows the proportion of those

(continued from previous page)
Education for Science and Engineering," background paper, forthcoming. Movement into and out of specific nonscience fields fell outside the purview of this report. However, the data revealed that, within this category, a single major—business—engaged an increasing proportion of of all students: 10 percent at sophomore year, 13 percent at senior year, and 15 percent in college. No other single major matched the growing appeal of this field.

Figure 1-2.—Persistence In, Entry Into, and Exit From Natural Science and Engineering By 1982 High School Graduates Planning Natural Science or Engineering Majors, 1980-84

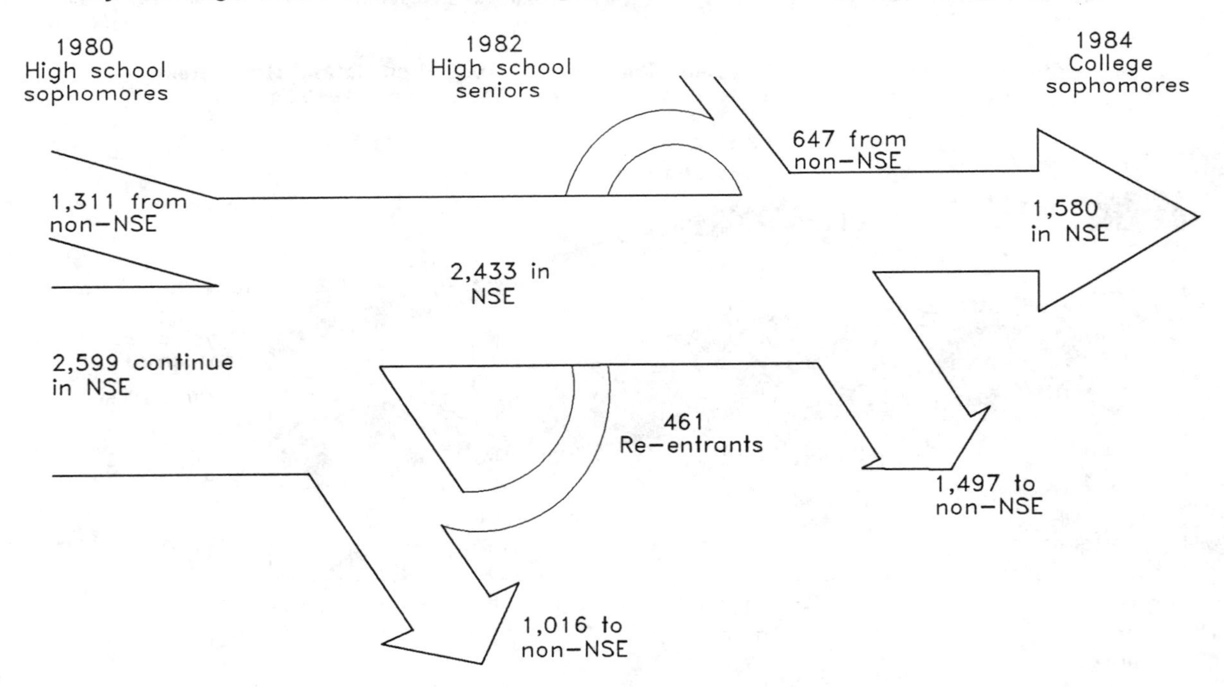

NOTE: This pipeline traces those students who, at some point, planned to major in natural science or engineering (NSE), out of a nationally representative sample of high school graduates (n = 10,739). "Re-entrants" chose NSE as high school sophomores, "left" NSE as high school seniors, but chose an NSE major in college. Only 300 students, or less than 10%, stayed with the same field within NSE at all three time points; the majority of NSE students changed field preferences *within* NSE at least once.

SOURCE: Valerie E. Lee, "Identifying Potential Scientists and Engineers: An Analysis of the High School-College Transition. Report 1: Descriptive Analysis of the High School Class of 1982," OTA contractor report, July 20, 1987, pp. 29-35. Based on data from U.S. Department of Education, High School and Beyond survey.

college sophomores who were intending to major in natural science and engineering, and had already expressed interest in these field categories at the two earlier time points. A surprisingly small number of students persisted in the same field category at all three time points—about 10 percent in the case of physical sciences and mathematics, and computer and information sciences, and about 25 percent for each of the other two field categories. Students planning engineering majors appear to have been the most persistent, since 60 percent of those declaring this intention during

Figure 1-3.—Planned Major in High School of College Students Majoring in Natural Science and Engineering, by Field, 1980-84

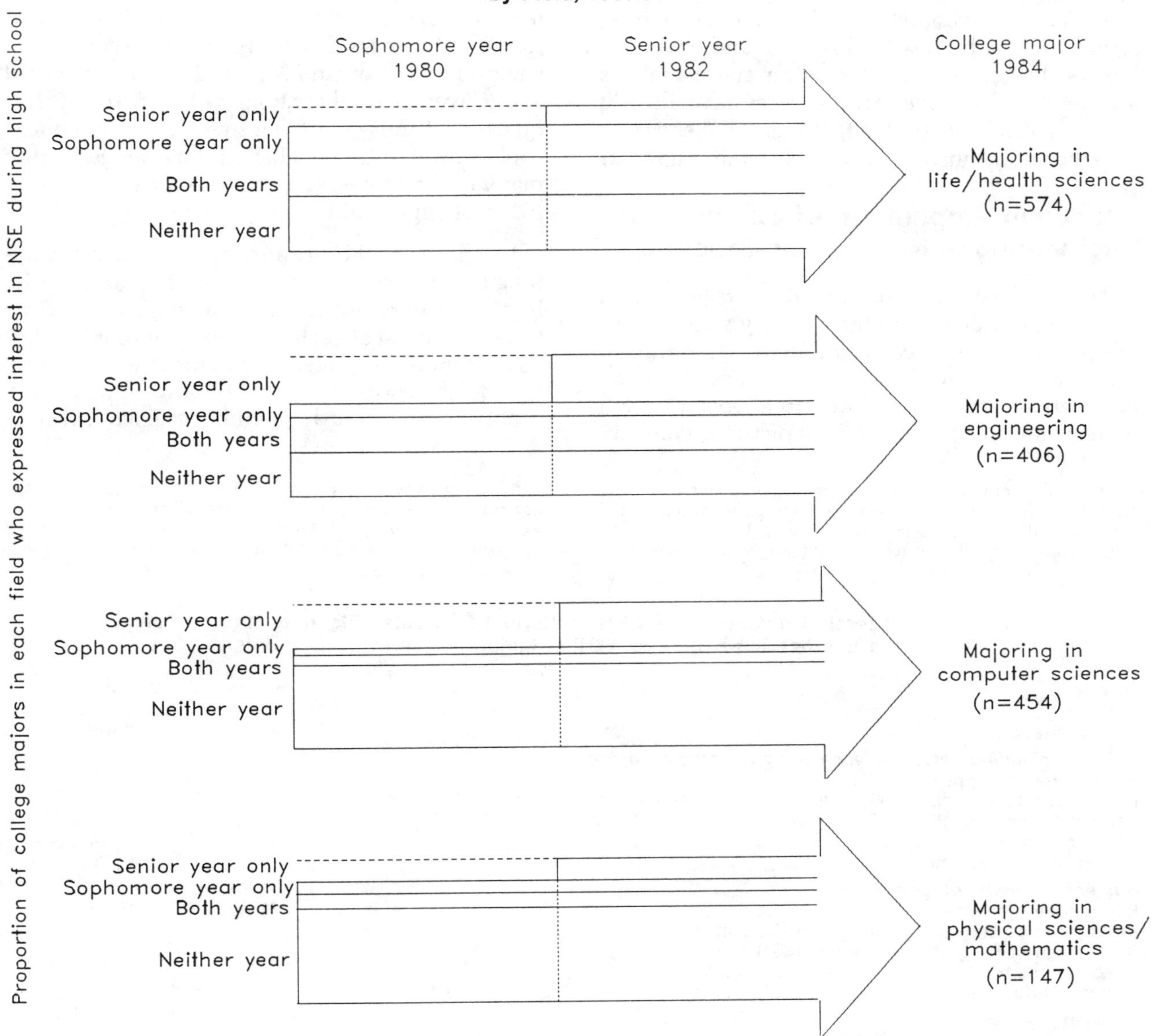

NOTE: This figure presents the high school history of college students majoring in natural science and engineering (NSE), showing when they expressed plans to major in their chosen college field. A large proportion of college NSE majors did not plan to major in their chosen field in high school; however, most planned to major in some NSE field. Based on a cohort of students who were high school sophomores in 1980.

SOURCE: Valerie E. Lee, "Identifying Potential Scientists and Engineers: An Analysis of the High School-College Transition. Report 1: Descriptive Analysis of the High School Class of 1982," OTA contractor report, July 20, 1987, p. 35. Based on data from U.S. Department of Education, High School and Beyond survey.

their high school senior year (and 34 percent of those in the sophomore year of high school) stayed with their plans.[12]

OTA's analysis of the HS&B survey shows that natural science and engineering attract some new adherents both in the later years of high school and the early years of college. The die is not cast in the early stages of the educational process; some students (approximately equal in number to those already in the high school science and engineering pool) enter that pool long after many analysts assume that definitive career choices have already been made. The interest is there; the challenge facing educational institutions is to capitalize upon it.

Academic Preparation of Science and Engineering v. Nonscience Students

The challenge of preparing future scientists and engineers is much more than simply sparking interest in students; it calls equally for preparation through coursework and a willingness to bring new entrants to the pipeline "up to speed." Data on new entrants reveal a mixed picture: many are very well prepared, but others take nontraditional routes and thus require extra help in mathematics and science courses.

Table 1-1 shows some of the characteristics of students planning natural science and engineering majors at the three survey time points. It also shows that the proportion of this group that scored above average on the HS&B achievement test[13] increased at each time point. About two-thirds of those students interested in science and engineering majors in 10th and 12th grades scored above average on these tests, but more than three-quarters of those planning such majors as college sophomores did so. This finding suggests that many of the new entrants to the pipeline are likely to be of high ability.

In addition, the proportion of students planning natural science and engineering majors who had been enrolled in the academic curriculum track increased at each time point until it reached 75 percent in the college sophomore year. Nevertheless, the corollary of this finding—that 25 percent of those seriously planning science and engi-

[12]It is important to note that this figure identifies only those who persisted in the same field category (and not those who persisted in science), although the data indicate few field differences in the numbers of students who entered each field category from nonscience majors.

[13]This test score is a composite of scores in reading, vocabulary, and mathematics from tests designed especially for the High School and Beyond survey and administered to the students when they were high school sophomores (in 1980). It has been highly correlated with other achievement tests.

Table 1-1.—Academic Characteristics of High School Graduates Planning Natural Science and Engineering Majors and Other Majors, at Three Time Points

Characteristic	1980 high school sophomores	1982 high school seniors	1984 college sophomores
Students planning natural science and engineering majors:			
Percentage of sample	24	23	15
Percentage scoring above 50% on HS&B achievement test	65	69	79
Percentage scoring above 75% on HS&B achievement test	39	44	53
Academic track	60	64	74
General and vocational tracks	40	36	26
Students planning other majors:			
Percentage of sample	42	38	44
Percentage scoring above 50% on HS&B achievement test	63	66	67
Percentage scoring above 75% on HS&B achievement test	36	37	37
Academic track	55	56	60
General and vocational tracks	45	44	40
Undecided major	22	21	3
No college plans	11	18	38

KEY: HS&B=High School and Beyond survey.

SOURCE: Valerie Lee, "Identifying Potential Scientists and Engineers: An Analysis of the High School-College Transition," OTA contractor report, 1987, based on the High School and Beyond survey.

neering majors enrolled in the vocational and general tracks in high school (with presumably less access to college-preparatory courses in mathematics and science)—indicates that the science and engineering pipeline contains some latecomers. Significant numbers of students are active participants in the college segment of the science and engineering pipeline without two of the traditional credentials of a future scientist and engineer: high ability manifested early on and academic track preparation.[14]

In a separate analysis (see table 1-2) of students who entered the pipeline from nonscience fields, either in their high school senior or college sophomore years, OTA found that these students, on average, had lower scores on achievement tests than did those who persisted in science at all three time points. The "in-migrants" to science and engineering majors had taken fewer mathematics and science courses and were more likely to be Black, Hispanic, or female than their "determined" science peers. Nevertheless, 70 percent of the in-migrants had been in the academic curriculum track and had high school and college grade point averages (GPAs) comparable to those who persisted in science and engineering throughout.

In a further comparison of those who switched from a nonscience to a science field during their last 2 years of high school with those who persisted in a nonscience field (see table 1-3), statistically significant differences were found in the course-taking patterns of the two groups. Immigrants to the science and engineering pipeline were more likely to have taken algebra II, calculus, chemistry, physics, or biology than their nonscience peers, and subsequently recorded higher GPAs in mathematics and science courses. Still, they were on average less well prepared than those who stayed with science plans from their high school sophomore to their college sophomore years.

The analysis of the high school class of 1982 illuminates several findings that demand rethinking of how the science and engineering pool forms. Taken together, these findings lend support to the recent observation of the National Academy of

[14]Nevertheless, the High School and Beyond data for 1980 high school sophomores had not yet followed through to college graduation, and so cannot be used to estimate what proportion of this "nontraditional" group succeeded in earning science and engineering baccalaureates.

Table 1-2.—Comparison of Students Who Persisted in Natural Science and Engineering With Those Who Entered These Fields, From High School Sophomore to College Sophomore Years, 1980-84

Characteristic	Persisted in same field N = 298	Persisted in NSE, but switched fields N = 277	Entered NSE from a nonscience field N = 1,004
Demographic characteristics:			
Percent Black	7	9	15
Percent Hispanic	5	9	10
Percent female	47	30	48
High school experiences:			
Number of math courses taken	3.1	3.0	2.7
Number of science courses taken	3.5	3.3	3.0
GPA in math courses	2.7	2.8	2.7
Score on HS&B Achievement Test[a]	58.5	58.3	55.0
Score on mathematics portion of SAT/ACT tests[b]	516	541	500
College experiences:			
College GPA	2.8	2.8	2.8
Percent attained college sophomore status by 1984	90	76	75

[a]On HS&B Achievement Test, mean score = 50, standard deviation = 10.
[b]Score is presented on same scale as the mathematics portion of the SAT; where students had taken the ACT mathematics test, their score was converted to an equivalent score on the SAT scale.
KEY: GPA = grade point average.
HS&B = High School and Beyond survey.
SAT/ACT = Scholastic Aptitude Test/American College Testing program.
NSE = natural science and engineering.

SOURCE: Valerie Lee, "Identifying Potential Scientists and Engineers: An Analysis of the High School-College Transition," OTA contractor report, 1987, based on the High School and Beyond survey.

Table 1-3.—Comparison of Students Who Persisted in Nonscience Interest With Those Who Entered a Natural Science and Engineering Major, From High School Sophomore Through College Senior Years, 1980-84

Characteristic	Persisted with a nonscience field N = 2,337	Entered a natural science and engineering major N = 799
Percent female	65	53[b]
Percent Black	8	11[b]
Percent Hispanic	9	10
Score on HS&B achievement test[a]	53.8	54.0
Percent in academic track	61	63
Mathematics GPA	2.3	2.5[b]
Science GPA	2.5	2.6[b]
Courses taken (percentage with 1 year or more):		
Mathematics		
Algebra 1	71	70
Geometry	55	59
Algebra 2	35	44[b]
Trigonometry	21	25[b]
Calculus	4	13[b]
Computer programming	4	5[b]
Science		
Biology 1	54	56
Advanced biology	15	21[b]
Chemistry 1	27	39[b]
Advanced chemistry	3	7[b]
Physics 1	11	26[b]
Advanced physics	1	3[b]

[a]On HS&B Achievement Test, mean score = 50, standard deviation = 10.
[b]Indicates that the difference between the two groups was statistically significant at $p < 0.05$.
NOTE: Percentages are rounded to the nearest whole number.
KEY: GPA = grade point average.
HS&B = High School and Beyond survey.

SOURCE: Valerie Lee, "Identifying Potential Scientists and Engineers: An Analysis of the High School-College Transition," OTA contractor report, 1987, based on the High School and Beyond survey.

Science's Government-University-Industry Research Roundtable:

> There are no magic one or two points in a student's life that are crucial to career choice. At every educational and developmental stage factors come into play that shape and reshape the occupational direction a student is taking. Moreover, the influences affecting different groups vary.[15]

[15]Government-University-Industry Research Roundtable, *Nurturing Science and Engineering Talent* (Washington, DC: National Academy Press, 1987), p. 34.

INTEREST AND QUALITY OF SCIENCE- AND ENGINEERING-BOUND STUDENTS

Interest in science and engineering, clearly, is not enough. The ultimate health of the science and engineering work force also depends on another key factor—the quality of students.[16] Science and engineering majors have traditionally had above average GPAs, college admission test scores, achievement test scores, and other markers of quality.

[16]There is little agreement on how the "quality" of high school students should be measured. Achievement test scores are only one indicator. Just as important are students' understanding of the process of science, attitudes toward science and engineering, language skills, and learning how to learn more effectively. Because there are no consistent data on these latter attributes, achievement test scores are here used as proxy for quality.

In recent years, somewhat fewer college freshmen with "A" or "A–" high school GPAs have chosen science and engineering majors, while increasing numbers name preprofessional and business majors. Yet the proportion of science- and engineering-bound students who score above 650 (of a maximum of 800) on the Scholastic Aptitude Test (SAT) mathematics test[17] has increased somewhat between 1975 and 1984. About 44 percent of those who score above the 90th percentile on the SAT mathematics test say they plan science and engineering majors. The average score of all those scoring above the 90th percentile on the SAT mathematics test increased from 623 to 642 in the last 5 years.[18]

It is widely believed by college educators that the quality of high school students who are planning science and engineering majors may be declining compared to their predecessors. While this belief has probably been held by all teachers who try to transmit knowledge to their juniors, there is little evidence to support it. Although SAT scores are an imperfect measure of the quality of students, the average SAT score of all students planning science and engineering majors declined between 1975 and 1983 (parallel to the decline in scores of the entire population of test-takers during the period from 1963 to 1981). The SAT scores of this group, however, have risen somewhat since 1983. The sources of increases and decreases in SAT scores provoke complicated and controversial debates in educational assessment, but there is some consensus that about one-half of the long-term decline during the 1960s and 1970s in the SAT scores was due to changes in the composition of the population of students taking the test. The remaining decline has been attributed to decreased emphasis on academic subjects by schools, and social factors. The recent increases are even less well understood.[19]

Overall interest in science and engineering appears to have increased since the time of the major longitudinal study centering on the high school class of 1972.[20] Since that time, there have been considerable shifts among fields within the science and engineering majors, often in response to employment markets. For example, the late-1970s and early-1980s saw a rapid increase in interest in engineering and computer science majors (freshman interest and college enrollment in engineering approximately doubled during that time), but some decline of interest in physical science majors.

International Comparisons

There is also current concern that America's best students are of inferior quality compared with their peers in other countries. Two recent international comparisons of achievement scores largely support this concern, but do not definitively explain the causes of these differences (although they suggest the curriculum as a culprit). In particular, interpretation of these data is complicated by major differences in the structure of education in different countries. (More detail on the mathematics and science educational systems of other countries is found in app. B.)

International comparison data are available from the Second International Math Study (SIMS) from 1981 to 1982, and the Second International Science Study (SISS) from 1983 to 1986.[21] These

[17]However, many science- and engineering-bound students take the American College Testing program test instead of the Scholastic Aptitude Test. Data on the science- and engineering-bound among the American College Testing program takers are not available.

[18]In particular, there are indications that highly talented white males, a traditional source of scientists and engineers, are increasingly being attracted to these majors. Mechanisms for increasing this group's participation need to be devised as well as those to increase the participation of women and minorities. See Office of Technology Assessment, op. cit., footnote 11. Also see National Science Board, *Science and Engineering Indicators 1987* (Washington, DC: U.S. Government Printing Office, 1987), pp. 24-25, app. table 1-7.

[19]National Science Board, *Science Indicators: The 1985 Report* (Washington, DC: U.S. Government Printing Office, 1985), p. 128. National Science Board, op. cit., footnote 18, p. 22. See also U.S. Congress, Congressional Budget Office, *Educational Achievement: Explanations and Implications of Recent Trends* (Washington, DC: U.S. Government Printing Office, August 1987).

[20]T.L. Hilton and V.E. Lee, "Student Interest and Persistence in Science: Changes in the Educational Pipeline in the Last Decade," *Journal of Higher Education*, vol. 59, September/October 1988, pp. 510-526.

[21]These studies were conducted under the auspices of the International Association for the Evaluation of Educational Achievement, a nongovernmental voluntary association of educational researchers. See International Association for the Evaluation of Educational Achievement, *Science Achievement in Seventeen Countries: A Preliminary Report* (Oxford, England: Pergamon Press, 1988); Willard J. Jacobson et al., *The Second IEA Science Study—U.S.*, revised edition (New York, NY: Columbia University Teachers College, September 1987); Curtis C. McKnight et al., *The Underachieving Curriculum: Assessing U.S. School Mathematics From an International*

(continued on next page)

data indicate that the performance in most science and mathematics subjects of U.S. science- and engineering-bound students is inferior to that of their counterparts in many other countries, including Japan, Hong Kong, England and Wales, and Sweden. One finding from the science studies is that the proportion of each nation's cohort of 18-year-olds who take college-preparatory science courses is apparently smaller in the United States than in other countries.[22]

Data from the first international mathematics and science studies (done in 1964 and 1970, respectively) indicate that the United States lagged behind other nations even then. For example, in the first international science study, students in Australia, Belgium, England and Wales, Finland, France, the Federal Republic of Germany, the Netherlands, Scotland, and Sweden scored above their peers in the United States (Japan did not participate). And, in the similar mathematics study, students in the United States scored lower than all of the above countries as well as Japan. Because there were few common test items between the first and second tests, and because data on the demographic and other characteristics of the groups tested were not collected at both time points, it is difficult to determine reliably from these studies whether achievement in the United States (or in any other country) has improved or declined. These studies suggest that, compared with other countries, the United States has fared, and continues to fare, poorly in the mathematical and scientific preparation of its future work force.

Interest in Science and Engineering Among Females and Minorities

Females and the members of some racial and ethnic minorities are represented in most fields of science and engineering in numbers far below their shares of the total population, a difference that emerges well before high school.[23]

Female interest in science and engineering is concentrated in the life and health sciences, and less so in more quantitative fields such as engineering and the physical sciences and mathematics (figure 1-4). Black and Hispanic high school seniors are about one-half as likely to be interested in careers in the physical sciences and mathematics as whites, and Blacks are about one-half as likely to be interested in engineering (figure 1-5).[24] Some of this difference may be due to the fact that Blacks and Hispanics are less likely to go on to college than whites.

Why are females and minorities less likely than males and whites, respectively, to major in science or engineering? What can be done about it? Discussion of these issues arouses vigorous debate, which stems in part from deeply embedded social attitudes and expectations about the roles and contributions of females and racial and ethnic minorities to American society, the professions, and specifically to the science and engineering work force.[25]

Differences Between the Sexes in Interest in Science and Engineering

There are considerable differences by sex in the number of students interested in science and engineering fields. Of those interested in life and health

(continued from previous page)
Perspective (Champaign, IL: Stipes Publishing Co., January 1987), pp. 22-30; F. Joe Crosswhite et al., *Second International Mathematics Study: Summary Report for the United States* (Washington, DC: National Center for Education Statistics, May 1985), pp. 4, 51, 61-68, 70-74; Robert Rothman, "Foreigners Outpace American Students in Science," *Education Week*, Apr. 29, 1987, p. 7; Wayne Riddle, Congressional Research Service, "Comparison of the Achievement of American Elementary and Secondary Students With Those Abroad—The Examinations Sponsored by the International Association for the Evaluation of Educational Achievement (IEA)," 86-683 EPW, June 30, 1986.

[22]For example, in these data, only 1 percent of American high school students are reportedly enrolled in chemistry and physics, but this refers only to seniors in high school who had taken second year chemistry and physics, not all students who took these courses (Richard M. Berry, personal communication, August 1988). In Canada, the numbers are 25 and 19 percent for chemistry and physics, respectively, and, in Japan, 16 and 11 percent, respectively. Note, however, that these enrollment data for the United States paint a considerably more pessimistic picture than earlier data on course-taking patterns has revealed (see ch. 2), and, due to small sample sizes, are subject to considerable uncertainty.

[23]For an overview, see Office of Technology Assessment, op. cit., footnote 4, chs. 2 and 3.

[24]Lee, op. cit., footnote 9.

[25]See, for example, Sandra Harding and Jean F. Barr (eds.), *Sex and Scientific Inquiry* (Chicago, IL: University of Chicago Press, 1987); Willie Pearson, Jr. and H. Kenneth Bechtel (eds.), *Education and the Coloring of American Science* (New Brunswick, NJ: Rutgers University Press, forthcoming); Marlaine E. Lockheed et al., *Sex and Ethnic Differences in Middle School Mathematics, Science, and Computer Science: What Do We Know?* (Princeton, NJ: Educational Testing Service, May 1985).

Figure 1-4.—Interest of 1982 High School Graduates in Natural Science and Engineering, by Sex, 1980-84

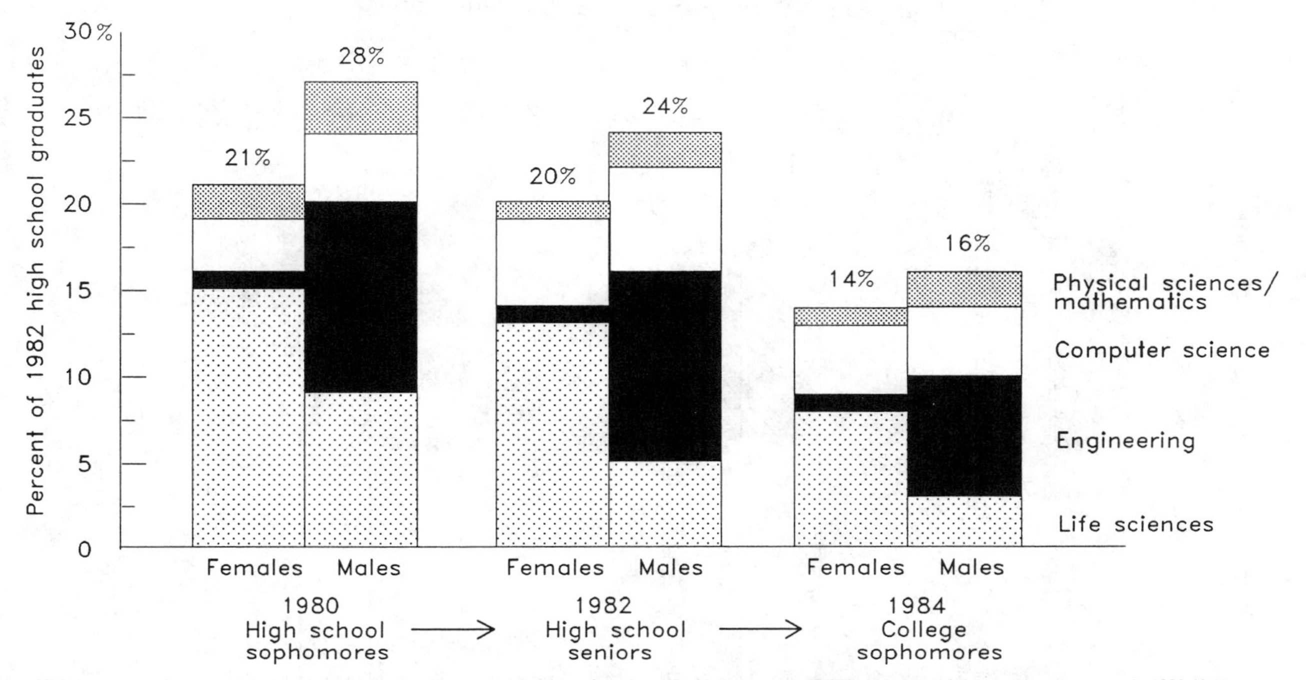

NOTE: This is a nationally representative sample of high school graduates (female n=5,466 and male n=5,273) who were high school sophomores in 1980. If the sample is restricted to an increasingly select group, of only those high school graduates who plan to go or have not ruled out going to college, the percent interested in natural science and engineering increases: 33% of male and 22% of female high school sophomores, 31% of male and 23% of female high school seniors, and 28% of male and 21% of female college sophomores. (The number of college-bound students decreases with time because some students planning college do not attend or drop out.)

SOURCE: Valerie E. Lee, "Identifying Potential Scientists and Engineers: An Analysis of the High School-College Transition. Report 1: Descriptive Analysis of the High School Class of 1982," OTA contractor report, July 20, 1987, p. 23. Based on data from U.S. Department of Education, High School & Beyond survey.

sciences, about one-half are males and one-half females. Many fewer females, however, are interested in fields with a significant mathematical component, such as physics, chemistry, and engineering. Interestingly, at the baccalaureate level, females are well represented in mathematics itself.

Males and females appear to differ strongly in their interest in highly quantitative sciences. Somewhat more males than females enroll in high school courses leading to these fields, but not so many more as to explain the size of the difference in interest between the sexes. Females also tend to score lower on mathematics achievement tests, even when allowances are made for the fewer courses they take compared to males. The exact causes of these differences in interest, course-taking, and achievement test scores have not been determined and remain a controversial subject for research.[26]

Many researchers believe that the differences are primarily or totally caused by the differential treatment that boys and girls receive from birth. Parents, friends, teachers, and counselors, it is argued, encourage males to be interested in mathematics and science and discourage females. Over time, females come to feel less confident than males about mathematics, come to believe that they do not have mathematical "talent," study mathematics less intensively, and hence score lower on achievement tests. Interest in this "environmental" hypothesis has led researchers to try to ascertain whether sex differences in interest or test scores can be related to these factors.[27]

[26]For a recent overview, see Valerie E. Lee, "When and Why Girls 'Leak' Out of High School Mathematics: A Closer Look," presented at the annual meeting of the American Educational Research Association, San Francisco, CA, April 1986 (unpublished paper available from ERIC). Research that proclaims the biological inferiority of females in mathematical (especially spatial) domains has been seriously questioned (discussed below).

[27]For example, see Patricia B. Campbell, "What's a Nice Girl Like You Doing in a Math Class?" *Phi Delta Kappan*, vol. 67, No. 7, March 1986, pp. 516-520. According to a recent national survey,

(continued on next page)

Figure 1-5.—Interest in Natural Science and Engineering by College-Bound 1982 High School Graduates, by Race/Ethnicity, 1980-84

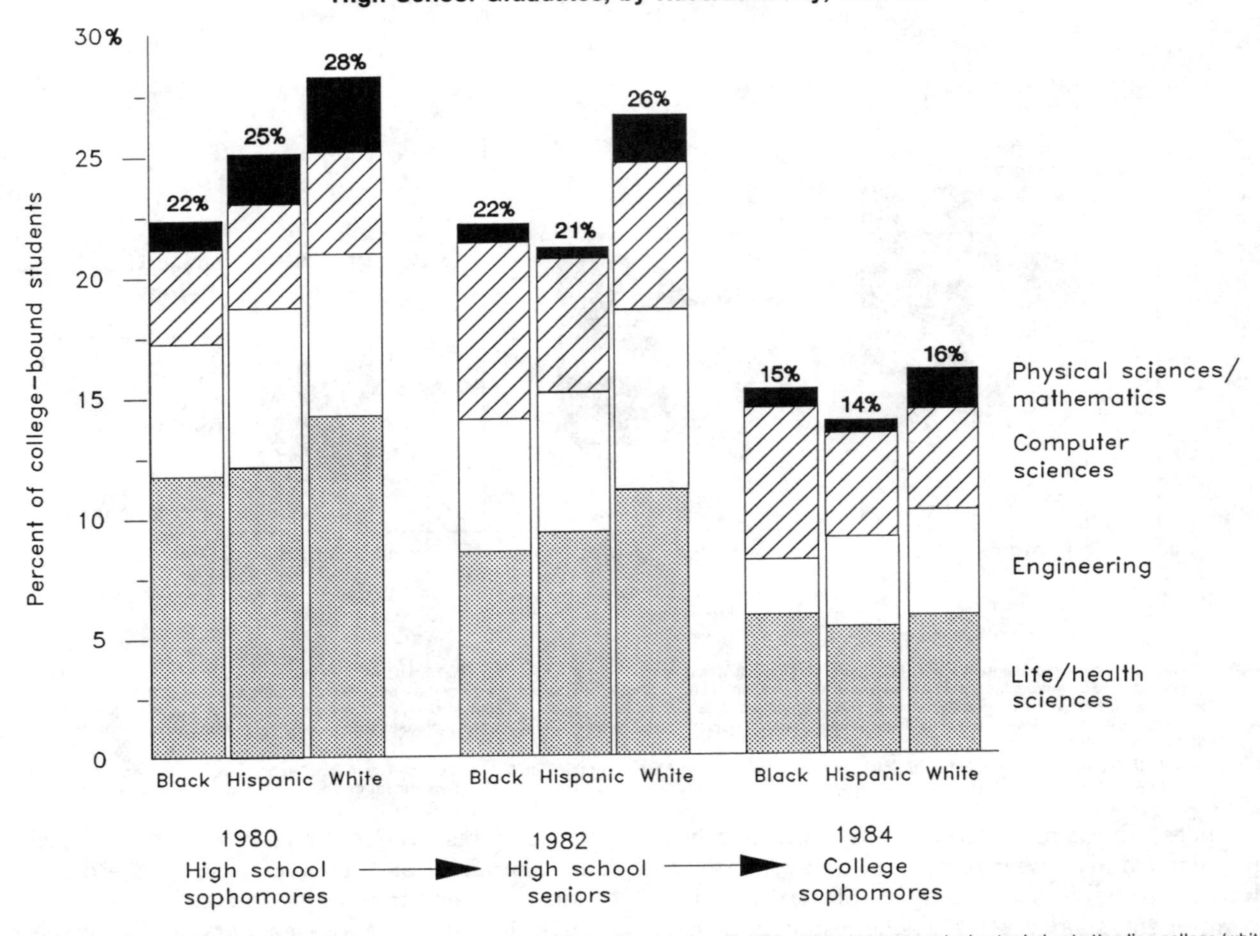

NOTE: The sample is limited to those high school graduates who, as high school sophomores in 1980, planned to go to or had not ruled out attending college (white n=7,541; Black n=1,321; Hispanic n=760). If the sample is restricted even further to those students who stay in the college pipeline after high school sophomore year, the percent interested in natural science and engineering increases; among college sophomores, to 28% of Blacks (n=718), 24% of Hispanics (n=434), and 23% of whites (n=5,206).

SOURCE: Valerie E. Lee, "Identifying Potential Scientists and Engineers: An Analysis of the High School-College Transition. Report 1: Descriptive Analysis of the High School Class of 1982," OTA contractor report, July 20, 1987, pp. 24-26. Based on data from U.S. Department of Education, High School and Beyond survey.

Others maintain that the differences are much more pervasive than can be explained by differential patterns of treatment and, therefore, that physiological differences between the sexes are at work. At the beginning of the decade, one study suggested that differences in the structure and function of the brain allow males to visualize spatial relationships better than females.[28] This hypothesis was based on analysis of scores on the mathematics portion of the SAT (a test designed for 11th and 12th graders) achieved by samples of highly talented 8-year-old males and females.[29] This hypothesis has been roundly criti-

(continued from previous page)

few students agree that mathematics is a subject more for boys than girls. In 1986, 6 percent of 13-year-olds and 3 percent of 17-year-olds agreed with this statement, although each of these percentages had increased from 2.5 and 2.2 percent, respectively, in 1978. See John A. Dossey et al., *The Mathematics Report Card: Are We Measuring Up? Trends and Achievement Based on the 1986 National Assessment* (Princeton, NJ: Educational Testing Service, June 1988), pp. 98-99.

[28]C.P. Benbow and J.C. Stanley, "Sex Differences in Mathematical Ability: Fact or Artifact?" *Science*, vol. 210, 1980, pp. 1262-1264; C.P. Benbow and J.C. Stanley, "Sex Differences in Mathematical Reasoning Ability: More Facts," *Science*, vol. 222, Dec. 2, 1983, pp. 1029-1031.

[29]A representative criticism is found in A.M. Pallas and K.A. Alexander, "Sex Differences in Quantitative SAT Performance: New

cized by many researchers on several counts and is not widely accepted.

It is unlikely that the controversy over the *origin* of gender-related differences in demonstrated mathematical ability will be resolved any time soon, as so many different factors must be controlled in studies making male-female comparisons. OTA concludes that effective steps can be taken to encourage females to enter science and engineering without detailed knowledge of the reasons for sex differences in mathematics achievement and for interest in science and engineering.

There is no evidence that the *rate* of learning of mathematics by males and females is different. *If* there are differences in the preparation for, orientation to, or talents of males and females in science, they can be remedied. Among such remedies are programs to sensitize parents, teachers, and counselors to their conscious and unconscious differential treatments of boys and girls.[30] Schools, and especially guidance counselors, can help significantly by encouraging females who do well in mathematics to take advanced mathematics courses. Schools should also encourage females to participate fully in hands-on scientific experiments. The encouragement of females to pursue science and engineering careers must counter continuing and pervasive, albeit decreasing, discrimination against females in the science and engineering work force, as indicated by lower salaries for new graduates and fewer females in tenured faculty positions.

Evidence on the Differential Coursework Hypothesis," *American Educational Research Journal*, vol. 20, No. 2, 1983, pp. 165-182. For a review of studies and instructive methodological commentary on the debate, see Susan F. Chipman and Veronica G. Thomas, "The Participation of Women and Minorities in Mathematical, Scientific, and Technical Fields," *Review of Research in Education*, Ernst Z. Rothkopf (ed.) (Washington, DC: American Educational Research Association, 1987), pp. 403-409.

[30]Myra Pollack Sadker and David Miller Sadker, *Sex Equity Handbook for Schools* (New York, NY: Longman, 1982); Jane Butler Kahle and Marsha Lakes Matyas, "Equitable Science and Mathematics Education: A Discrepancy Model," *Women: Their Underrepresentation and Career Differentials in Science and Engineering*, Linda S. Dix (ed.) (Washington, DC: National Academy Press, 1987), pp. 5-41.

Photo credit: William Mills, Montgomery County Public Schools

Females and minorities are an undertapped source of scientists and engineers.

Reasons Why Minorities Are Not Well Represented in Science

Blacks show interest in science and engineering and, in particular, in careers in engineering, mathematics, and computer science.[31] The challenge is to convert this interest into well-prepared future scientists and engineers.

Development of interest and talent in science and engineering by Blacks is stultified by their relatively lower average socioeconomic status and more limited access to courses that prepare them for science and engineering careers. Minorities also sense hostility from the largely white science and engineering work force and develop low expectations for themselves in science and mathematics courses. For some, sadly, success in academic study is scorned by their minority peers as "acting white." Larger proportions of minorities drop out of high school than do whites, reducing the potential talent pool for science and engineering. And far fewer Black males than Black females prepare for college study, a pattern which is increasingly common as Black males favor military

[31]Jane Butler Kahle, "Can Positive Minority Attitudes Lead to Achievement Gains in Science? Analysis of the 1977 National Assessment of Educational Progress, Attitudes Toward Science," *Science Education*, vol. 66, No. 4, 1982, pp. 539-546; Mary Budd Rowe, "Why Don't Blacks Pick Science?" *The Science Teacher*, vol. 44, 1977, pp. 34-35. Lee, op. cit., footnote 9. Data from the most recent National Assessment of Educational Progress show that 48 percent of Blacks in grade three said that they would like to work at a job using mathematics v. 38 percent of white students. Students in grades 7 and 11 were asked a different question, whether they expected to work in an area that requires mathematics. In grade 7, 46 percent of whites and 39 percent of Blacks said yes and in grade 11, 45 percent of whites and 51 percent of Blacks said yes. See Dossey et al., op. cit., footnote 27, pp. 95-100.

service or become convinced that even a college degree is no guarantee of a good job.[32]

Schools, school districts, and States need to do much more to help minorities gain access to science and engineering careers. Teachers need to be sensitized to minority concerns and to involve all students in mathematics and science experiences, such as laboratory experiments. Schools need to develop better guidance counseling, and inculcate higher expectations for minorities among counselors, teachers, and students. Schools and school districts also need to improve their course offerings and ensure that all students have access to the preparatory courses leading to a science and engineering degree. And States and school districts need to ensure that schools with high minority populations receive a fair share of financial, teaching, and equipment resources, given that such schools are often in poor areas.[33]

Undertaking such reforms will be difficult for an education system that changes very slowly. In any event, such actions still will not overcome wider societal pressures that minorities believe deter them from science and engineering careers. Specific intervention programs (discussed in ch. 5) are needed to overcome these deterrents.

[32]Signithia Fordham, "Recklessness as a Factor in Black Student's School Success: Pragmatic Strategy or Pyrrhic Victory," *Harvard Educational Review*, vol. 58, No. 1, February 1988, pp. 54-84; *ERIC/SMEAC Information Bulletin*, "A Review of the Literature on Blacks and Mathematics," No. 1, 1985.

[33]A veritable flood of successful interventions with minority students, in schools and out, is detailed in Lisbeth B. Schorr, *Within Our Reach: Breaking the Cycle of Disadvantage* (New York, NY: Anchor Press Doubleday, 1988), especially chs. 8-12.

SCHOOLS AS TALENT SCOUTS

The goals of excellence and equity both depend on taking full advantage of the Nation's talent. Doing so depends on having schools act in large measure as talent scouts. Instead, the schools have often acted as curricular traffic cops, encouraging the obviously talented, and culling out those who do not display the conventional signs of ability and drive at an early age. Too many students "never play the game" of science. (See box 1-A.) The result is a waste of talent.

Similar wastes of nonscientific talent undoubtedly take place. The importance of high school preparation in science and mathematics to future success in these careers, though, makes waste particularly serious in these fields. In the future, schools will need to cast their nets wider in identifying potential scientists and engineers, going beyond the standard model of talent. The growing ethnic and cultural diversity of young Americans makes this task both more challenging and more important. The remainder of this report will detail the steps schools and communities might take to meet this national need.

Box 1-A.—Never Playing the Game

Learning science—its theories, its parameters, its context—can be likened to learning all the rules of a sport—the facilities needed for playing, the scoring, the timing, the uniforms. To prepare to play, one must develop physical skills by means of strenuous exercise and conditioning. Such skill development can span great time periods and demand much energy, commitment, and sacrifice. But, most potential players are willing to devote whatever time and work are needed to succeed, because once they reach the playing field, their effort will be rewarded.

In typical science teaching, we ignore the lessons we might learn from sports. We pronounce science a fantastic game—that all should learn to play it. We spend years teaching background material, laws, rules, classification schemes, and verifications (disciplines) of the basic game. We plan activities for our students designed to develop in them specific skills that the best scientists seem to possess and use. We believe that proficiency with these skills is an important part of an education in science. It is as if we were developing conditioning exercises to train our students for the science they may actually do at a future time.

Unfortunately, however, our students rarely get to play—rarely get to do real science, to investigate a problem that they have identified, to formulate possible explanations, to devise tests for individual explanations. Instead, school science means 13 years of learning the rules of the game, practicing verification-type labs, learning the accepted explanations developed by others, and the special vocabulary and the procedures others have devised and used.

If potential athletes had to wait 13 years before playing a single scrimmage, a single set, a single quarter, how many would be clamoring to be involved? How many would do the pull-ups and the sit-ups? How many would learn the rules if there were no rewards—until college—for those who had practiced enough to play?

We expect much in science education! Could one of our problems be too much promise of what science is really like at a date much too far removed from the rigor and practice science demands? Thirteen years of preparation is a long time to wait before finding out whether a sport (or career) is as satisfying as one's parents and teachers suggest it will be.

To prepare for the game of science for 13 years without even an opportunity to play is a problem! Like athletes, science students may need to play the game frequently, to use the information and skills they possess, and to encounter a real need for more background and more skills. Such an entree to real science in school could result in more students wanting to know and wanting to practice the necessary skills. Now, we lose too many students with only the promise that the background information and skills we require them to practice will be useful.

Paul Brandwein asserts that most students never have a single experience with real science throughout their whole schooling. He has written that we would have a revolution on our hands if every student had but one experience with real science each year he or she is in school. Are we ready for such a revolution? Can we afford not to clamor for it?

To spend 13 years preparing for a game, but never once to play it, is too much for anyone. Teachers and students alike are more motivated when they experience real questions, follow up on real curiosity, and experience the thrill of creating explanations and the fun of testing their own ideas. Real science must become a central focus in the courses we call science—across the entire K-12 curriculum.

SOURCE: Quoted from Robert E. Yager, "Never Playing the Game," *The Science Teacher*, September 1988, p. 77.

Chapter 2

Formal Mathematics and Science Education

Photo credit: Neldine Nichols, Wisconsin Department of Public Instruction

CONTENTS

Boxes

Figures

Tables

Chapter 2

Formal Mathematics and Science Education

There has been widespread concern among scientists and educators alike over the failure of the instructional programs in the primary and secondary schools to arouse greater interest and understanding of the scientific disciplines. . . . There is general agreement that much of the science taught in schools today does not reflect the current state of knowledge nor does it represent the best possible choice of material for instructional purposes.

Hearings before the Committee on Appropriations, U.S. Senate, 1958

For most people, science education begins and ends in school mathematics and science classes. Science and mathematics education, however, cannot be analyzed in isolation from the overall national system of K-12 education. The national pattern of schooling mixes diversity of control and decisionmaking with a surprising uniformity of organization and goals.

THE NATION'S SCHOOLS

About 75 percent of each birth cohort now graduates from high school. Elementary and secondary education takes place in over 100,000 schools, employing 2.5 million teachers and enrolling about 45 million students.[1] It costs $170 billion per year, about $4,000 per student—or 4 percent of the annual gross national product.[2] Public education, including higher education, is the single largest component of State spending. Nationally, the States' contribution to the cost of public school education is about 49 percent of the total; 45 percent comes from local sources and the remaining 6 percent from the Federal Government. The Federal Government's contribution peaked at about 10 percent of the total from 1978 to 1980.[3] (See figure 2-1.) The balance among local, State, and Federal funding varies significantly from State to State, however. New Hampshire has the highest proportion of local funding, at 90 percent; Hawaii has the highest proportion of State

Figure 2-1.—Funding of Elementary and Secondary Education, by Source, 1980-87 (constant 1987 dollars)

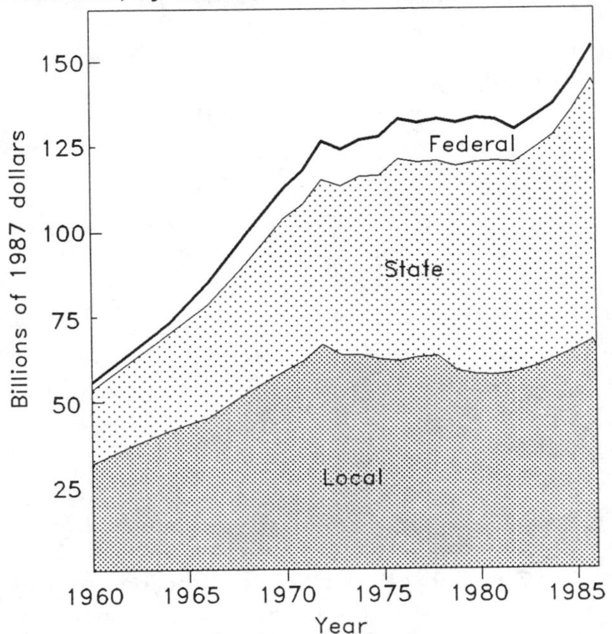

SOURCE: U.S. Department of Education, National Center for Education Statistics, *Digest of Education Statistics 1987* (Washington, DC: 1987), p. 107; and unpublished data.

[1]U.S. Department of Education, Office of Educational Research and Improvement, Center for Education Statistics, *Digest of Education Statistics 1987* (Washington, DC: U.S. Government Printing Office, May 1987), tables 3, 4, 5, 59. Data on the number of public schools is from 1983-84, that on private schools from 1980-81. The number of school districts is from 1983-84. Student enrollment data is an estimate for fall 1986, number of teachers are estimated full-time equivalent excluding support staff for fall 1986.

[2]Ibid., table 21. Data are estimates for 1986-87.

[3]Ibid., table 93 (preliminary data for 1984-85).

funding, at 89 percent; and Mississippi has the highest proportion of Federal funding, at 16.5 percent.[4] The bulk of the cost of education is in providing buildings and paying salaries for teachers and other staff. A tiny proportion is spent on instructional materials, such as textbooks and laboratory equipment.[5] Private schools are funded from tuition charged to students, although they also receive tax benefits and participate in some Federal aid programs. They enroll about 10 percent of students.[6] (See figure 2-2.)

[4]Hawaii is the only State to have only one school district. That is, the public schools are, in effect, State run rather than school board run.

[5]For example, over one-half of public expenditures on elementary and secondary schools in 1979-80 went to "instruction," primarily teacher salaries. U.S. Department of Education, op. cit., footnote 1, table 96. According to data from the American Association of Publishers, the average school district spent $34 on instructional materials per pupil in 1986, of a total spending of $4,000 per pupil, just under 1 percent. Also see Harriet Tyson-Bernstein, testimony before the House Subcommittee on Science, Research, and Technology of the Committee on Science, Space, and Technology, Mar. 22, 1988.

[6]The system of public schools is paralleled by an extensive system of private education for which parents pay tuition (in addition to the taxes that support public schools). The majority of private schools are of religious foundation.

Figure 2-2.—Public and Private School Enrollments and Revenues, by Source, 1985-86

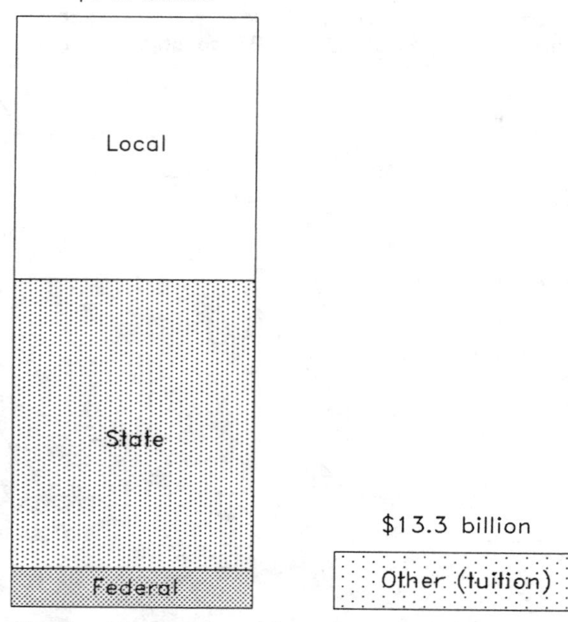

SOURCE: U.S. Department of Education, National Center for Education Statistics.

Although control over public education is in the hands of the States and the school districts, schooling across the Nation displays remarkable uniformity.[7] Compulsory education generally begins at age 6, in grade 1; for those who persist, free public schooling ends in grade 12 at age 18. There are three commonly used school classifications: elementary (kindergarten to grade 5, 6, or 7); middle or junior high (grade 6, 7, or 8 to grade 8 or 9); and high (grades 9 or 10 to 12). Schools award grades on a scale that normally stretches from 1 to 4, curricula and course titles are fairly uniform, and a student's grade point average is nationally recognized as a measure of the student's progress.

The Nation's 84,300 public schools are controlled and run by autonomous school districts, subject to State laws and systems of organization. These school districts are of very unequal size: the largest 1 percent educate 26 percent of students, while the smallest 43 percent educate only 4 percent of students. The trend is toward consolidation of smaller districts, and the number of school districts has fallen in 50 years from 120,000 to slightly less than 16,000 today.[8] Most school districts, although geographically coextensive with local governmental units such as counties and cities, raise their own funds through local taxes and bond issues, and are run by locally elected school boards. In other districts, funding and control are the responsibilities of counties and cities.

Just as education nationally displays diversity *and* uniformity, so it is with mathematics and science education. The National Science Teachers Association estimates that at least 50 percent (8.3

[7]See, for example, Barbara Benham Tye, "The Deep Structure of Schooling," *Phi Delta Kappan*, December 1987, pp. 281-284.

[8]U.S. Department of Education, op. cit., footnote 1, tables 5, 59, 60, 62. Data on the distribution of students among school districts are from fall 1983. Data on the number of public schools are for 1983-84. An often overlooked element of local control in education is the composition of school boards. Ninety-five percent of the 97,000 school board members in the United States are elected. They make policy on everything from school lunch menus to textbook adoption affecting 40 million students. Jeremiah Floyd, National School Boards Association, remarks at Workshop on Strengthening and Enlarging the Pool of Minority High School Graduates Prepared for Science and Engineering Career Options, Congressional Black Caucus Braintrust on Science and Technology, Washington, DC, Sept. 16, 1988.

million) of the high school population enrolled in a science class in 1986.[9] In 1981-82, 78 percent of high school students (9.9 million) were enrolled in mathematics courses.[10] To provide these courses in 1985-86 required a work force of about 100,000 science teachers and 173,000 mathematics teachers, who together made up about 30 percent of the secondary school teaching force.[11]

[9]National Science Teachers Association, *Survey Analysis of U.S. Public and Private High Schools: 1985-86* (Washington, DC: March 1987), p. 5. A 1981-82 analysis indicated about the same number of enrollments in high school science courses, Evaluation Technologies, Inc., *A Trend Study of High School Offerings and Enrollments: 1972-73 and 1981-82*, NCES 84-224 (Washington, DC: National Center for Education Statistics, December 1984), p. 17.

[10]Evaluation Technologies, Inc., op. cit., footnote 9, p. 16.

[11]National Science Teachers Association, op. cit., footnote 9, p. 2; National Education Association, *Status of the American Public School Teacher, 1985-86* (West Haven, CT: 1987), p. 11, table 17. (The National Education Association estimates that 11 percent and 4.5 percent of elementary school teachers that specialize in a subject area teach mathematics and science, respectively.)

COMPONENTS OF MATHEMATICS AND SCIENCE CURRICULA

For many children, the content of mathematics and science classes and the way these subjects are taught critically affect their interest and later participation in science and engineering. The effectiveness of different teaching techniques is addressed in the next chapter, but this section reviews the controversies over the typical American mathematics and science curriculum, the alleged dullness of many science textbooks, and the extent to which greater use of educational technology, such as computers, could improve the teaching of mathematics and science.[12]

Many mathematics and science educators are critical of the quality of the mathematics and science curricula in use in most schools. They see the curricula as slow-moving, as failing to draw links between scientific and mathematical knowledge and real-world problems, and as relatively impervious to reform.[13]

[12]Similar criticisms are made of the entire K-12 curriculum that, according to many observers, fails to draw links between separate courses or between its content and the workings of the outside world. A recent Carnegie Foundation report called for ". . . a kind of peacetime Manhattan Project on the school curriculum. . . ." which would ". . . design, for optional State use, courses in language, history, science and the like and . . . propose ways to link school content to the realities of life." The Carnegie Foundation for the Advancement of Teaching, *Report Card on School Reform: The Teachers Speak* (Washington, DC: 1988), p. 3. See also Fred M. Newman, "Can Depth Replace Coverage in the High School Curriculum," *Phi Delta Kappan*, January 1988, p. 345.

[13]Attempts were made to improve mathematics and science curricula in the 1960s via several projects funded by the National Science Foundation. These projects did have some success, and have affected overall curricula in these subjects in particular by stressing the use of practical experiments. Details on these federally funded programs appear in ch. 6.

Typical Mathematics and Science Curricula

The mathematics and science curricula in use in schools are fairly standardized, due to the widespread use of the Scholastic Aptitude Test and American College Testing program for college admissions, college admission requirements, the workings of the school textbook market (which ensures considerable uniformity of content), and the need to accommodate students who transfer from one school to another. Nevertheless, there are important differences in the problems of curricula between mathematics and science.[14]

In mathematics, grades one to seven are devoted to learning and practicing routine arithmetical exercises. In grade seven, the more advanced students may move on to take courses that prepare them for algebra, but the most commonly offered class is "general mathematics." The most able and motivated take algebra in grade eight (if their school offers it), and there is some evidence that the number of algebra classes in grades seven to nine is rising.[15] In higher grades, students go on to courses in advanced algebra (also known as algebra II), geometry, trigonometry, and either precalculus or calculus (where these are offered). (See figure 2-3.) About 10 percent of each cohort

[14]See, on science curricula generally, Audrey B. Champagne and Leslie E. Hornig (eds.), *The Science Curriculum: The Report of the 1986 National Forum for School Science* (Washington, DC: American Association for the Advancement of Science, 1987).

[15]Iris R. Weiss, *Report of the 1985-86 National Survey of Science and Mathematics Education* (Research Triangle Park, NC: Research Triangle Institute, November 1987), pp. 24-25.

Figure 2-3.—Typical Course Progression in Mathematics and Science

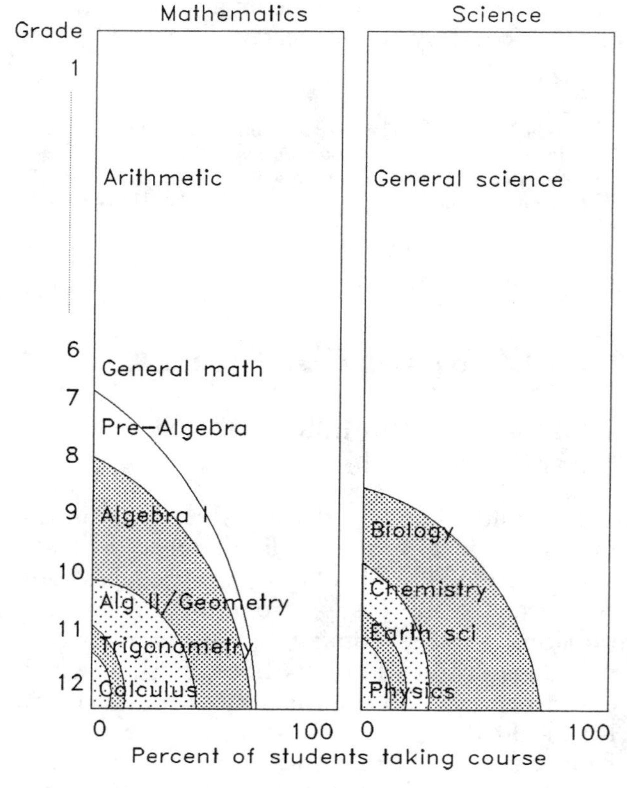

SOURCE: Office of Technology Assessment, 1988.

of high school graduates persist in mathematics long enough to take a calculus course.

The abstract and undemanding pace of the mathematics curriculum at all levels has been widely criticized recently. Many mathematics educators believe that there is too much emphasis on arithmetical drill and practice, as opposed to an emphasis on understanding mathematical concepts and applications. Many highly able students in mathematics, likely future science and engineering majors, report that the average curriculum proceeds at too slow a pace.[16] The fact that handheld calculators—in widespread use in everyday life nationwide for over a decade—are still rarely found or used in mathematics (and science) classrooms is particularly criticized. A 1985 survey found that about one-half of mathematics and science classes in grades 10-12 used calculators at some point, but the proportion was much lower in earlier grades (14 percent in the case of grade K-6 mathematics classes).[17] Another survey found that almost all students reported that either they or their family owned a calculator, but less than one-quarter said that their schools had calculators for use in mathematics classes.[18]

Many mathematics educators also note that the mathematics curriculum was largely left out of earlier reform efforts, which concentrated mainly on science. The experiments in "new math" left a bad taste in many mouths, and there still remains deep suspicion among teachers, parents, and school boards of attempts to develop a "new" mathematics curriculum. Nevertheless, mathematics educators are using international comparisons of curricula, teaching practices, and achievement test scores to demonstrate the inferiority of many American mathematics curricula.[19] The National Council of Teachers of Mathematics is undertaking a consultation program to develop curriculum and assessment standards for school mathematics, and the Mathematical Sciences Education Board, part of the National Research Council, is launching a broad-based reform program designed, in part, to alert parents, school boards, and teachers of the need to improve school mathematics. (See box 2-A.) Particular efforts are also being made to improve the teaching of calculus—increasingly taught in high school and, by many accounts, appallingly taught at the college level.[20]

[16]Benjamin S. Bloom (ed.), *Developing Talent in Young People* (New York, NY: Ballantine, 1985), pp. 303-311; Lynn Arthur Steen, "Mathematics Education: A Predictor of Scientific Competitiveness," *Science*, vol. 237, July 17, 1987, pp. 251, 252, 302.

[17]Weiss, op. cit., footnote 15, tables 24 and 31, pp. 48, 56.

[18]John A. Dossey et al., *The Mathematics Report Card: Are We Measuring Up? Trends and Achievement Based on the 1986 National Assessment* (Princeton, NJ: Educational Testing Service, June 1988), p. 79.

[19]Curtis C. McKnight et al., *The Underachieving Curriculum: Assessing U.S. School Mathematics From an International Perspective* (Champaign, IL: Stipes Publishing Co., January 1987); and Robert Rothman, "In 'Bold Stroke,' Chicago to Issue Calculators to All 4th-8th Graders," *Education Week*, Oct. 14, 1987.

[20]National Council of Teachers of Mathematics, "Curriculum and Evaluation Standards for School Mathematics: Working Draft," unpublished manuscript, October 1987; Robert Rothman, "Math Group Sets New 'Vision' for Curriculum," *Education Week*, Nov. 11, 1987, p. 5; and Lynn Arthur Steen (ed.), *Calculus for a New Century: A Pump, Not a Filter* (Washington, DC: Mathematical Association of America, 1988).

Box 2-A.—The Special Role of the Science and Engineering Research Communities in Mathematics and Science Reforms

Science is an active endeavor. In laboratories and research facilities all over the world, scientists, engineers, and technicians form research communities. These communities provide the colleagues, verification, peer review, and legitimacy of the research enterprise. Members of these communities also prepare future generations of researchers. Many members of research communities have an active interest in science education, even at the precollege level, for they are concerned both with communicating the joy and beauty of science and with nurturing future waves of talent. Yet, despite the obvious potential linkages between mathematics and science education and the practice of science and engineering, there is still a continuing need to forge those linkages.

Teachers and educators often lament that research communities are not actively helping them and remain indifferent to the early preparation of future talent. To an extent, there will always be friction between these groups; many scientists disdain educators, whom they see as merely mass-produced and disseminating yesterday's knowledge. Deeper mutual understanding are needed.

Two major initiatives are under way to bridge communities and enhance education: the American Association for the Advancement of Science's Project 2061 and the Mathematical Sciences Education Board, based at the National Research Council. Project 2061 has brought together a group of prominent scientists to reform science and mathematics education for the next 70 years—when Halley's Comet will next return to Earth. The Project has three phases: content identification, educational formulation, and educational transformation. In the first phase, the group is attempting to define the science, technology, and mathematics that all high school graduates should know. In the second phase, these "goals for learning" will transmute into educational guidelines for curricula, school organization, teacher training and support, and testing methods. Finally, in the third phase:

> . . . the strategies and mechanisms needed to reform American schooling in the light of the intellectual framework of Phase I and the educational guidelines of Phase II will be established and monitored. This phase will have to be a highly cooperative, nationwide effort which will mobilize resources, monitor progress, and, in general, provide direction and continuity of effort.[1]

The Mathematical Sciences Education Board was established in October 1985 with the aim of launching a major reform effort for elementary, secondary, and undergraduate teaching of mathematics, focusing in particular on curricula. The primary purpose of the Board is ". . . to provide a continuing national overview and assessment capability for mathematics education." The Board works under the auspices of the National Research Council, and on it sit mathematicians, mathematics educators, and people familiar with schools and school systems. The Board is working on increasing public understanding of school mathematics issues, raising national expectations for mathematics teaching and learning, and reaching a consensus on goals and education for future mathematics teaching. It is planning ways to help States and school districts improve their performance in mathematics.[2]

[1]F. James Rutherford et al., "Project 2061: Education for a Changing Future," *The Science Curriculum: The Report of the 1986 National Forum for School Science*, Audrey B. Champagne and Leslie E. Hornig (eds.) (Washington, DC: American Association for the Advancement of Science, 1987), pp. 61-65.

[2]Lynn Arthur Steen (ed.), Calculus for a New Century: A Pump, Not a Filter (Washington, DC: Mathematical Association of America, 1988).

In science, until grade eight, most students take general science courses. In grade nine, upon transfer to high school, they typically take general science, biology, then chemistry, earth science, and finally physics. About 20 percent of high school graduates persist in science long enough to take physics.[21] In many schools, these courses are taught separately, and teachers fail to draw the links and contrasts among the different science disciplines. The typical order of their presentation—biology, chemistry, physics—is deeply ingrained in the culture of American school science, but some teachers feel that other arrangements would be better, and would draw more effective links

[21]In 1987, 20.1 percent of high school graduates earned minimal credits in physics, up from 13.9 percent in 1982. Westat, "1987 High School Transcript Study," unpublished tabulations for the "Nation at Risk Update Study," May 1988, p. 110.

with mathematics courses. (For comparison, see figure 2-4, which outlines the order of courses followed by a science magnet school in which this traditional order is reversed.

Problems With Textbooks

Most mathematics and science in schools is taught with the aid of textbooks, which are an area of considerable controversy (more so in science than in mathematics).[22] Many scientists and science educators are deeply critical of typical science textbooks, which they say concentrate largely on defining terms without explaining their origin and the scientific concepts that they describe. Science, they say, ends up being presented

[22]A 1985-86 survey of teachers indicated that over 90 percent of mathematics and science classes in grades seven to nine used a published textbook or program, a proportion which has remained level since 1977. About two-thirds of elementary science classes use them, as do over 90 percent of elementary mathematics classes. The survey also found that most teachers claimed to cover at least 75 percent of the book that they used. See Weiss, op. cit., footnote 15, tables 14 and 19. A similar survey of students found that three-quarters of students in grade 7 and 11 reported using mathematics textbooks in classes daily, and only 4 or 5 percent, respectively, "never" used textbooks in class. In grade three classes, more use is made of workbooks or ditto sheets than textbooks. See Dossey et al., op. cit., footnote 18, p. 78.

Figure 2-4.—Curriculum of a Mathematics/Science/Computer Science Magnet Program for the Class of 1991

Year	Mathematics	Science	Seminar	Computer Science
Grade 9	***Course sequence***			
1st semester	Magnet Geometry A&B ↓ (1 credit)	Advanced Science 1 Physics (1 credit)	Research and Experimentation Techniques for Problem Solving 1 including: Probability and Statistics Research Methods (½ credit)	Fundamentals of Computer Science A (½ credit)
2nd semester	Magnet Functions A&B (1 credit) or	Advanced Science 2 Chemistry (1 credit)		Fundamentals of Computer Science B (½ credit)
Grade 10				
1st semester	Magnet Precalculus A,B,C ↓ (1½ credits)	Advanced Science 3 Earth Science (½ credit)	Research and Experimentation Techniques for Problem Solving 2 (½ credit)	Algorithms and Data Structures A (½ credit)
2nd semester	Analysis I A&B ↓ (1 credit)	Advanced Science 4 Biology (1 credit)	(½ credit)	Algorithms and Data Structures B (½ credit)
Grade 11 1st semester 2nd semester	Analysis II (½ credit) or Linear Algebra or Discrete Mathematics or Excursion Topics in Math or Guided Research, Internship Cooperatives, University courses, etc.	Advanced mini-courses, research, internships, university courses, special topics, cooperatives, etc. (variable credit) Examples: Climatology, Tectonics, Metallurgy, Cellular physiology, Biomedical seminar, Thermodynamics, Optics, Cooperatives	Research and Experimentation Techniques for Problem Solving 3 (1 credit)	Advanced topics in semester and mini-courses, university study, special topic sessions, projects (variable credit) Examples: Analysis of Algorithms, Graphics, Survey of Languages, Computer Architecture & Organization, Game Theory
Grade 12 1st semester 2nd semester			Guided senior project involving research and/or development across discipline lines (1 credit)	

NOTE: Elective courses for grades 11, 12 include options like: Advanced Placement courses. Game Theory, Topology, Mathematical Programming, Abstract Algebra, Cooperative Languages, Robotics, Computer Architecture, Systems Design, Organic Chemistry, Quantitative analysis, Astrophysics, Plant Physiology, Behavior and Brain Chemistry, Calculus in Biology/Ecology.

SOURCE: Montgomery Blair High School, Silver Spring, MD, September 1987.

as a monolith of unconnected and unchallengeable "facts," which are learned only by those students with extraordinary memories and with an overriding determination to pass the standardized tests of their ability to recall such definitions. (The need to address "facts," as well as their interpretation and construction, is discussed in box 2-B.) As a result, students find the textbooks boring. For example, one recent review by a science educator of a newly revised biology textbook notes:

> [This] product offers facts, pseudofacts and cliches in a matrix of rote sentences and plentiful pictures. It continually fails to integrate information, to explain concepts or to explain biology, and it is rich in absurdity. The writers reduce the topic of "scientific methods" to two paragraphs within a confusing passage on "The Origin of Life." . . . The book . . is attractive but scientifically meaningless.
>
> [The book] presents a frenetic display of facts—a smothering blanket of facts—and it will not inspire scientific thinking in any student or teacher. At most, it will impart an artificial and shallow sense of learning while it damages imagination and creativity.[23]

[23]National Center for Science Education, Inc., *Bookwatch*, vol. 1, No. 1, February 1988.

Box 2-B.—Learning About Science and Processes of Advancing Scientific Knowledge

Science is widely equated by the American public with a huge body of knowledge about physical and human processes that the scientific research enterprise has created. Many teachers similarly equate science with the mass of facts and material found in textbooks. Teachers report that their job is to cover as much of this material as they can and get their students to learn and memorize it. Many teachers say that successive editions of textbooks get bigger and more expensive, as more and more factual material is added. In fact, some teachers are very critical of many textbooks in current use because they feel that they convey a misleading impression of the reality of science and engineering. While scientists share the view that science is a large collection of facts, there are other important dimensions that science students must learn and appreciate to thrive in college-level study.[1]

Science is better viewed as a subject that embraces both a body of knowledge *and* the process by which that knowledge is developed and verified.[2] Future scientists and engineers need to know about what is already known and about how new knowledge can be created.[3] Most scientific knowledge rests on experiments that yield data. These data either confirm or fail to support hypotheses that scientists make about relationships in the physical world. (Rare exceptions are wholly theoretical fields, where physical experiments are intractable and knowledge is built up by theoretical analysis.) Thus knowledge of how experiments are conducted, what sources of error there might be in them, how potentially contradictory evidence is sought and evaluated, and how the results of experiments can be used—both within science and without—is the key to understanding the process of doing science.[4]

A further dimension to the study of science is that of its effects on society and its role in society and government. This dimension is collectively known as "science, technology, and society," and is being introduced into many school science classes.[5]

[1]As the late Roger Nichols, former director of the Boston Museum of Science, said about science: "We take a thing which is essentially a process and convert it into a reading exercise. It's no wonder that the overwhelming majority of children are turned off from science by the eighth grade." Gordon L. Nelson, "Playing 'Basketball' in the Smithsonian," *Chemical & Engineering News*, vol. 66, No. 18, May 2, 1988, p. 38.

[2]Historians and philosophers of science, increasingly in the company of scientists and sociologists of science, have generated a sizable literature in the last quarter-century on science as a *social* process replete with the drama of human emotions—competing egos, frustrating errors, and prizes won and lost. See, for example, Stephen Toulmin, "From Form to Function: Philosophy and History of Science in the 1950s and Now," *Daedalus*, vol. 106, summer 1977, pp. 143-162; Bruno Latour and Steve Woolgar, *Laboratory Life: The Social Construction of Scientific Facts* (Beverly Hills, CA: Sage Publications, 1979); and Karin Knorr-Cetina and Michael Mulkay, *Science Observed: Perspectives on the Social Study of Science* (London, England: Sage Publications, 1983).

[3]Mary Budd Rowe, "Science Education: A Framework for Decision-Makers," *Daedalus*, spring 1983, pp. 122-142. Related concerns about the competing pulls of depth v. breadth in the curriculum occur in other subjects. See Fred M. Newmann, "Can Depth Replace Coverage in the High School Curriculum?" *Phi Delta Kappan*, vol. 69, January 1988, pp. 345-348; and Robert E. Yager, "What's Wrong With School Science?" *Science Teacher*, January 1986, pp. 145-147.

[4]Educational Technology Center, *Making Sense of the Future* (Cambridge, MA: Harvard Graduate School of Education, January 1988).

[5]Newsletters that unite and inform elementary and secondary teachers of science, technology, and society include the *SSTS Reporter* (Science Through Science, Technology and Society) edited at Pennsylvania State University, the *Teachers Clearinghouse for Science and Society Education Newsletter* (sponsored by the Association of Teachers in Independent Schools, New York City), and the *Hawkhill Science Newsletter* (for Scientific Literacy), produced by Hawkhill Associates, Inc., Madison, WI.

The evidence is that both the economics of textbook publication and the politics of textbook selection result in "watered-down," poorly written, but attractive and fact-filled textbooks.[24] Some educators bitterly criticize the process of textbook "adoption" (approved for use statewide) used in 22 States. These States have a great deal of control over the content of the books. As a result, textbooks are designed to meet adoption criteria in the few key States such as Texas and California (in kindergarten to grade eight only) that guarantee the largest market if the book is approved. States that do not have textbook adoption mechanisms consequently have a more limited choice of textbooks, because publishers are reluctant to incur the cost of producing new volumes for these smaller markets.

Because of the pressure to include material that will satisfy textbook adoption committees and in order to outdo rival publishers, textbooks typically are large and heavy, profusely illustrated, are often printed in full color, but are surprisingly uniform in content. The books typically increase in size and weight with each edition, becoming more expensive, harder to carry, and more difficult for students to take home and study. The textbooks often include a huge quantity of material, in order to ensure that each State's recommended science curriculum is covered and that all interest groups are mollified. However, many important but controversial aspects of science, most notoriously the theory of evolution, may be omitted or given inadequate treatment.

[24]For criticism of the general textbook situation and suggestions for policy reform, see Harriet Tyson-Bernstein, *A Conspiracy of Good Intentions: America's Textbook Fiasco* (Washington, DC: The Council for Basic Education, 1988). Also see Tyson-Bernstein, op. cit., footnote 5; and Richard P. Feynman, *Surely You're Joking, Mr. Feynman: Adventures of a Curious Character* (New York, NY: Bantam Books, 1986), pp. 262-276.

Photo credit: William Mills, Montgomery County Public Schools

Formal education relies heavily on textbooks marketed by major publishing companies. Most educators find that these textbooks vary greatly in quality.

Interest groups lobby State textbook adoption committees to ensure that their own viewpoint is included, but the effect is that new text and pictures are added—material is rarely deleted. Ultimately, depth is sacrificed for breadth. And, because the adoption process typically involves an expert panel that quickly skims each volume, the textbooks often are designed to have key words in prominent places and be attractively packaged. The result is often textbooks that are a "lowest common denominator" of inoffensive facts, limited in conveying to students the process of constructing new scientific knowledge or the uses that are made of it. Some have described science textbooks as "glossaries masquerading as textbooks." Often, the "facts" themselves are old or entirely discredited. As a recent analysis of the process of textbook adoption notes:

> We have dozens of powerful ministries of education [States] issuing undisciplined lists of particulars that publishers must include in the textbooks. Since publishers must sell in as many jurisdictions as possible in order to turn a profit, their books must incorporate this melange of test-oriented trivia, pedagogical faddism and inconsistent social messages. . . . Under current selection procedures, those responsible for choosing the best among available books seem blind to the incoherence and unreadability of the book because they are merely ascertaining the *presence* of the required material, not its depth or clarity.[25]

In economic terms, the textbook market is quite concentrated, as indicated in table 2-1. For example, almost half of all elementary mathematics classes and 37 percent of all elementary science classes use one of the three most commonly used textbooks in these grades. Only a few textbook publishers supply much of the market. Yet a number of smaller publishers happily coexist along side

[25]Tyson-Bernstein, *A Conspiracy of Good Intentions*, op. cit., footnote 24, pp. 7, 109-110; also see National Science Board, *Science & Engineering Indicators, 1987* (Washington, DC: U.S. Government Printing Office, 1987), pp. 35-36.

Table 2-1.—Most Commonly Used Mathematics and Science Textbooks

Publisher	Title	Percentage of classes that use book
Science, grades K-6:		
Silver Burdett	*Science: Understanding Your Environment*	17
Merrill	*Accent on Science*	10
D.C. Heath	*Science*	10
Science, grades 7-9:		
Merrill	*Focus on Life Science*	9
Merrill	*Principles of Science*	8
Merrill	*Focus on Physical Science*	8
Science, grades 10-12:		
Holt, Rinehart, Winston	*Modern Biology*	14
Holt, Rinehart, Winston	*Modern Chemistry*	9
Merrill	*Chemistry: A Modern Course*	5
Mathematics, grades K-6:		
Addison-Wesley	*Mathematics in Our World*	16
D.C. Heath	*Mathematics*	15
Scott, Foresman	*Invitation to Mathematics*	7
Mathematics, grades 7-9:		
Houghton Mifflin	*Algebra: Structure and Method*	7
D.C. Heath	*Mathematics*	4
Scott, Foresman	*Mathematics Around Us*	4
Mathematics, grades 10-12:		
Houghton Mifflin	*Algebra: Structure and Method*	14
Houghton Mifflin	*Geometry*	8
Holt, Rinehart, Winston	*Algebra With Trigonometry*	2

SOURCE: Iris R. Weiss, *Report of the 1985-86 National Survey of Science and Mathematics Education* (Research Triangle Park, NC: Research Triangle Institute, November 1987), tables C.1 and C.2.

the major publishers, supplying either materials for parts of courses or entire texts.[26] Nevertheless, it is not a huge market since it is estimated that, in 1986, the total sales of instructional materials was equivalent to about $34 per student (only about 1 percent of the annual cost of education per student).[27]

Despite these gloomy assessments, a recent survey of mathematics and science teachers found that only a minority of them were concerned about textbook quality. When asked whether the poor quality of textbooks was a serious problem in their school, 11 percent of K-6 science teachers and 5 percent of grade 10-12 science teachers said yes. Fewer than 8 percent of mathematics teachers thought it was a problem. Between 15 and 25 percent of teachers thought that the quality of textbooks was somewhat of a problem. Factors such as large class sizes, inadequate access to computers, lack of funds for equipment and supplies, inadequate facilities, poor student reading abilities, and students' lack of interest were cited to be more serious problems. Indeed, teachers rated the organization, clarity, and reading level of textbooks favorably. Elementary teachers had more favorable ratings of textbooks than did secondary teachers.[28]

Textbooks pose several contradictions: most teachers seem (rightly or wrongly) to like the textbooks they use, many outside reviewers are skeptical of the scientific worth of many mathematics and science textbooks, there are apparently no overwhelming barriers to entry to the market, and powerful political and economic forces shape the dynamics of the textbook market. Devising "better" textbooks is not enough, for they must be adopted to be used and they are likely, on present

[26]A 1985-86 survey found that 10 publishers (Addison-Wesley; Harcourt Brace Jovanovich; D.C. Heath; Holt, Rinehart, Winston; Houghton Mifflin; Laidlaw; MacMillan; Merrill; Scott, Foresman; and Silver Burdett) accounted for at least three-quarters of all mathematics and science textbooks at all levels, and that two publishers accounted for almost one-half of each of elementary mathematics and science textbooks. See Weiss, op. cit., footnote 15, pp. 32-37.

[27]Tyson-Bernstein, op. cit., footnote 5, p. 13 and table II. Based on data from the Association of American Publishers, 1986.

[28]Weiss, op. cit., footnote 15, pp. 40-42 and tables 20, 21, and 71.

evidence, to be severely corrupted in the process. The best teachers have the ability to go beyond the material in textbooks, to provide supplementary material and examples, and to weave the concepts that the books try to explain into some coherent whole. But many teachers do not have the time, energy, or authority to use materials other than the approved texts.

Overall, the deficiencies, if any, in current mathematics and science textbooks stem from the divorce between the buyers and users of textbooks. Greater teacher involvement in textbook selection, a more courageous selection of members of textbook selection committees by States, and a greater participation by qualified scientists and engineers in the textbook adoption process will help, but not rectify, the problem.[29]

Use of Computers in Mathematics and Science Education

Computers offer new approaches for learning mathematics and science for all children, and help prepare students for college courses and technical careers that will demand familiarity with the technology.[30] If used well and imaginatively, computers can increase students' interest and improve learning, particularly for both the most and least advanced students. But the educational impact of computers is limited when the rest of the classroom environment stays the same. Conclusive research on effectiveness is meager.[31]

Computer technology and software are evolving. Educators are still discovering both positive and negative impacts on students, classrooms, and learning; how best to use the technologies to improve learning; and what support is needed in teacher training, curriculum modification, and research. Use of computers by teachers and students is still quite limited, although their availability and the equality of access enjoyed by different schools and students is improving. The use and future impact of computers depends on familiar features of the rest of the school system: curricula, time, quality of overall science and mathematics instruction, and, most of all, the comfort and competence of teachers with computers.

Nature and Extent of Use

The potential of computers is slowly being realized. Regular computer use is not extensive, although almost all schools now have at least some computers available. Primarily because of the small number of computers relative to students (the computer-to-student ratio in science classes is estimated to be 1:15 in middle school and ranging from 1:10 to 1:17 in high school, which is higher than the average in all subject areas), computers are most commonly used as infrequent enrichments rather than as an integral part of science teaching.[32] About one-third of high school mathematics teachers use computers in the classroom.[33] Nevertheless, much of this is occasional

[29]There is a Federal role, too. The National Science Foundation has issued a "publisher initiative" that outlines criteria for needed student assessment materials in baseline science development projects for the elementary grades and middle school. Since 1987, the National Science Foundation has funded seven "Troika" programs ". . . intended to encourage partnerships among publishers, school systems, and scientists/science educators for the purpose of developing and disseminating a number of competitive, high quality, alternative science programs for use in typical American elementary schools." See the National Science Foundation, Science and Engineering Education Directorate, Instructional Materials Development Program, "Publisher Initiative" and "The 'Troika' Program," unpublished documents, July 1988.

[30]This discussion centers on the now-familiar desktop personal computer or computer with keyboard and/or mouse input. Some schools have networked computers, or links with computers at other sites. Some schools, in particular science-intensive schools, have more powerful computers, computerized laboratory instrumentation, and computer-aided data processing equipment. Other information technologies, such as interactive videodiscs, are also powerful learning tools but they are less widespread than computers. Calculators, present in increasing numbers of classrooms, are having a greater effect than computers, because they reach many more students and are more readily linked in teachers' minds with existing curricula items, such as arithmetic. For more detail, see U.S. Congress, Office of Technology Assessment, *Power On! New Tools for Teaching and Learning*, OTA-SET-379 (Washington, DC: U.S. Government Printing Office, September 1988).

[31]Henry J. Becker, Center for Research on Elementary and Middle Schools, "The Impact of Computer Use on Children's Learning: What Research Has Shown and What It Has Not," unpublished manuscript, 1988.

[32]Sylvia Shafto and Joanne Capper, "Doing Science Together," *Teaching, Learning and Technology: A Digest of Research With Practical Implications*, vol. 1, No. 2, summer 1987, p. 2. A 1985 survey found that over 90 percent of schools had access to computers, but that only in about one-quarter of mathematics and science classes was this equipment readily available. Often, it is shared with other classes, or kept in special-purpose rooms that must be scheduled in advance. Weiss, op. cit., footnote 15, table 68.

[33]Becker found 17 percent for middle and secondary schools, while the National Science Teachers Association found 26 percent for all schools with a grade 12 (all secondary schools and a few middle schools). See Henry Becker, "1985 National Survey," *Instructional Uses of School Computers*, No. 4, June 1987; and National

use, and amounts to a small fraction of instructional time.[34] A 1985 survey noted that ". . . there is only the hint that secondary school science instruction might be profoundly affected by computers. The impact is largely still in the future."[35]

Computers are not used intensively in science classes. (See figure 2-5.) In secondary school, only 5 to 10 percent of computer use is for science; in elementary school, it is about 1 percent. There has been a slight decrease in the number of science programs available over the past 5 years, especially in chemistry and physics.

A unique application of computers in science is in the microcomputer-based laboratory (MBL), where computers can simulate experiments or process and display data obtained from simulated experiments.[36] (See box 2-C.) The use of science laboratories in science teaching has declined for many reasons, but computers can reduce some of the barriers to laboratory work, such as the rising cost of supplies, purchasing and maintaining equipment, concerns about safety hazards and liability, limited teacher competence in experimental work, the complexity of some experimental procedures, and the "one time—look quick" nature of many laboratories.[37] MBLs can also help children learn the process of science and research—hypothesis formation, testing, asking "what if" questions, gathering and analyzing data—at their own pace. Students using MBLs have shown greater understanding of basic principles and skills such as graphing data than students in regular laboratories.[38]

Figure 2-5.—Time Spent Using Computers in Mathematics and Science Classes, by Grade, in Minutes Per Week, 1985-86

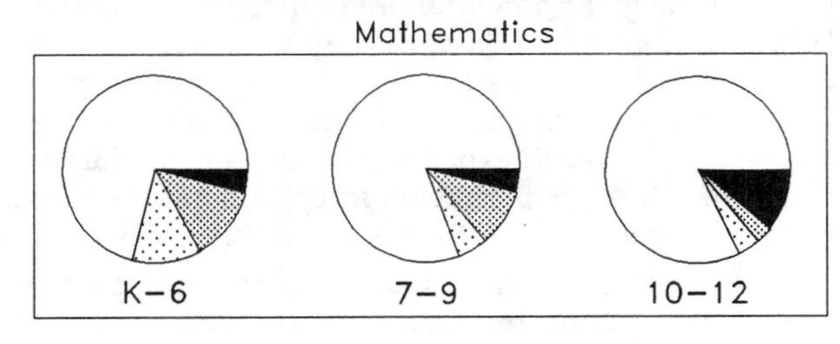

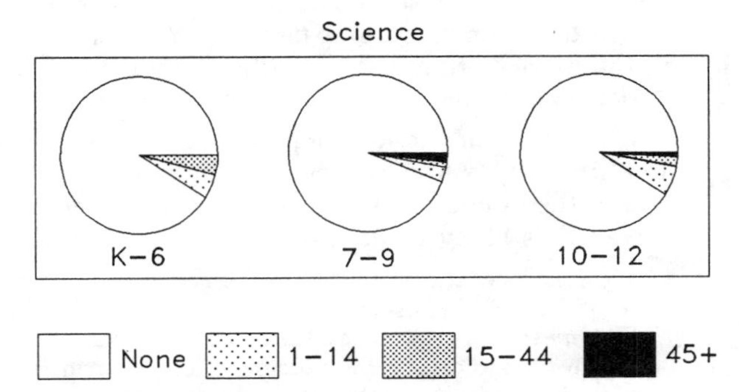

SOURCE: Iris R. Weiss, *Report of the 1985-86 National Survey of Science and Mathematics Education* (Research Triangle Park, NC: Research Triangle Institute, November 1987), p. 59.

Besides limited access to machines, other important barriers to more extensive use of computers are teachers' lack of familiarity with the technology and the lack of educational software (computer programs). Surveys indicate that comparatively few mathematics and science teachers have taken courses either in the instructional uses of computers or in computer programming. Only a bare majority of secondary mathematics teachers have taken either of these courses (table 2-2).

About one-half of educational software is devoted to topics in mathematics, science, and computer literacy. Mathematics was one of the first applications of educational computing, and con-

Science Teachers Association, "Survey Analysis of U.S. Public and Private High Schools: 1985-86," unpublished document, March 1987.

[34]A 1985-86 survey found that, in grades 10-12, 10 percent of mathematics classes and 5 percent of science classes used computers in their last lesson, but one-third of courses of each type used them at some point. Elementary students spend more time with computers than secondary students, but still two-thirds of K-6 mathematics classes and 85 percent of K-6 science classes report not using computers at all "last week." Weiss, op. cit., footnote 15, pp. 48, 58, 59, tables 24, 32, 33. Another recent survey has found computer access for learning mathematics to be relatively equitable across the sexes, races, and ethnicities, although high-ability students are more likely to report that they have access than are low-ability students. About one-third of high school juniors have taken a computer programming course. Dossey et al., op. cit., footnote 18, pp. 82-86.

[35]Becker, op. cit., footnote 33, p. 11.

[36]Shafto and Capper, op. cit., footnote 32, p. 1. However, computers are used much more widely in the science classroom than in the laboratory.

[37]Alan Lesgold, "Computer Resources for Learning, *Peabody Journal of Education*, vol. 62, No. 2, 1985, cited in Shafto and Capper, op. cit., footnote 32, p. 4. A 1985 survey found that the number of mathematics and science classes that had used "hands-on" activities in their most recent lesson had declined at all grade levels (with the sole exception in K-3 mathematics) since 1977. For example, while 53 percent of grade 10-12 science classes in 1977 reported using hands-on activities, in 1985 only 39 percent did. Weiss, op. cit., footnote 15, table 25, p. 49.

[38]Office of Technology Assessment, op. cit., footnote 30, ch. 5.

Box 2-C.—National Geographic Society (NGS) Kids Network: Students as Scientists

One great potential of computers is to allow children to do "real science"—to design and conduct experiments, collect data, display the results, try "what if" experiments, compare their results with those of other students, look for unusual results and patterns, and try to understand what it all means. Most school laboratories run "cookbook" experiments with predetermined outcomes. There are various ways computers are used to teach experimentation: to simulate experiments; to provide real-time, visual data analysis of regular "wet" experiments in microcomputer-based laboratories; and, in the NGS Kids Network, to use telecommunications to link children across the Nation in a real experiment.

The NGS Kids Network uses networked computers and telecommunications to link elementary school children around the United States. During a month-long 1987 pilot Kids Network project on acid rain, students in nine States did classroom work, collected rain samples, measured pH, and shared their results over a telecommunications network. A project coordinator analyzed the results and distributed them to all the participating classes. With the reach and immediacy of national data collection and analysis, the students could see their results, and those of all other students, overnight.

The chief impact of the NGS Kids Network, as with much computer use, seems to be in getting children, teachers, and parents excited and involved. Enthusiasm generally improves learning. Formal evaluation of the impact of the Kids Network and computers on understanding science concepts and methods, however, is likely to be less clear-cut.

> Results of the trial far exceeded expectations. The students puzzled over their measurements and wondered why Nebraska had such an unexpectedly high level of acid rain. (In fact, this finding is in accord with recent EPA measurements.) They learned about acids and bases, about dilution, pH scales, the range of accuracy of measuring indicators, the possible effects of acid rain, and latitude and longitude. Students became engrossed in finding out more about the communities where the other sites were located. They learned valuable lessons about the process of science, engaged in spirited discussions about scientific method, the validity of data, and the need for care in measurement. Students were highly motivated, bringing in newspaper clippings, collecting additional samples, and spontaneously writing to students at other sites and to public officials. Uniformly impressed by the unit, teachers enjoyed teaching it.[1]

Future NGS Kids Network units will be self-contained, 1-month packages of classroom activities. "Close to home" topics such as water and air pollution, food growing, solar energy, and weather forecasting have been chosen because they adapt well to distributed, national observations and data collection. One teacher commented:

> After they had brought in their samples of rain water and saw that they had gotten different results, they decided that they hadn't collected enough data. They made a judgement that they didn't have enough data to make an accurate conclusion. I see that as a very important step in understanding science. . . . They got the message that science is a collective enterprise as opposed to an individual one. That message came through very strongly.[2]

Special resource materials, teacher training, and user-friendly software have been developed to make the sophisticated units extremely easy for the novice student and teacher to use. To communicate results, all the teacher or student has to do is switch on the computer. Dissemination will be through the National Science Teachers Association, science centers, and the extensive National Geographic infrastructure.

The Kids Network is being developed by the Technical Education Research Centers and is funded by the National Science Foundation and the National Geographic Society. With the development costs thus supported, the costs to the schools can be kept low; telecommunications costs are a large and uncertain factor. Total development costs, including government and private grants and in-kind equipment and service donations, are estimated at a total of $5 million over 4 years.

[1]Robert Tinker, Technical Education Research Centers, "Network Science Arrives," *Hands On!* vol. 20, No. 1, winter 1987, pp. 1, 10-11.
[2]Ibid.

Table 2-2.—Courses in Computers Taken by Mathematics and Science Teachers

	Percentage of teachers that have taken	
	Instructional uses of computers	Computer programming
Mathematics teachers:		
Grades K-3	30	17
Grades 4-6	34	24
Grades 7-9	40	46
Grades 10-12	42	64
Science teachers:		
Grades K-3	31	11
Grades 4-6	37	21
Grades 7-9	33	33
Grades 10-12	30	33

SOURCE: Iris R. Weiss, *Report of the 1985-86 National Survey of Science and Mathematics Education* (Research Triangle Park, NC: Research Triangle Institute, November 1987), tables 39, 40, 41, and 44.

tinues to be the subject where students are most likely to encounter computers. More software is available for mathematics than for any other subject area, although most of it is for learning and practicing basic skills.[39] For example, interactive computer graphics can be powerful in helping children construct graphs and visualize algebraic and geometric functions.

Computer Impacts, Opportunities, and Needs

Computers offer the potential for individualized instruction. If carefully developed and used, computers and software can open doors to mathematics and science for students, particularly females and minorities, who traditionally have had limited interest or success in these courses. In many settings, however, computers can also reinforce existing patterns and stereotypes: boys tend to crowd girls away from computers, affluent children benefit from computers at home and more extensive access at school. Little is known as yet of the impact of different kinds and intensity of computer use on interest in, and preparation for, different college majors.

Computers also make it possible to offer courses that might not otherwise be available. Distance learning or packaged computer courses can enrich the schooling of advanced high school students or rural students in schools with limited course offerings. Likewise, familiarity with computers is becoming expected of incoming college students, particularly in science and engineering. If students do not have the opportunity to work with computers, computers may become yet another barrier to attainment of educational goals.

A pressing need is to help current science and mathematics teachers become comfortable using computers in the classroom and laboratory. Although most new teachers being trained are exposed to computers, many of them still report not being comfortable using them.[40] Technology training has unique aspects that distinguish it from other inservice training, in particular a need for special facilities and equipment. Teachers must have generous access to computers to use them effectively.

Continuing research on learning and evaluations of the effectiveness of computer-aided education are needed. Trials of different schools and learning structures would help in evaluating the strengths and weaknesses of computer technologies. The challenge is to measure the process of learning, and not just the content and outcomes of acquired knowledge. Another need is development of computer-integrated curricula, which builds the strengths of computers into curricula from the start rather than appending them to existing curricular frameworks.

The Federal Government has helped schools acquire computers, supported research on their uses and their integration with curricula, and to some extent augmented private sector development of hardware, software, and services.[41] Some Department of Education funds, although not specifically targeted to computers, have

[39] Ibid.

[40] Less than one-third of recent graduates feel prepared to teach with computers. See ibid., p. 98.

[41] Ibid., and Arthur S. Melmed and Robert A. Burnham, "New Information Technology Directions for American Education: Improving Science and Mathematics Education," report to the National Science Foundation, unpublished manuscript, December 1987.

helped schools acquire hardware and software. Title II of the Education for Economic Security Act of 1984 (see ch. 6) has supported teacher training. The National Science Foundation has been instrumental in software development, networking among schools, and teacher training. Federal influence has been small compared to that from equipment manufacturers or vendors (among whom Apple has been prominent) and private foundations.[42] States are active in the movement to improve basic skills (including mathematics and science literacy), in which computers are playing an increasing role.

[42]The Federal Government could promote software development. Melmed and Burnham, op. cit., footnote 41, pp. 12-13, suggest that the Federal Government fund four mathematics and four science curricula to meet the goal of a science and mathematics course each year of high school. A rough estimate is that such development would cost about $2 million ($1 to $4 million) per course. Development should include review of old and existing curricula. Distribution and maintenance might total 25 percent of development costs, but could be recovered through a school user fee. Trials would need to be several years long.

VARIATION AMONG SCHOOLS

As any parent knows, schools vary in a myriad of ways; their location, control, and funding may affect their children's progress. For example, there is some evidence that private Catholic schools are especially effective at channeling their students toward academic college education, owing to the personal attention and high expectations they give their students.[43]

OTA did not analyze data on the special features of mathematics and science instruction in private schools compared with public schools. Rather, it considered the contrasts among urban, suburban, and rural public schools, as related to their respective socioeconomic settings and expectations of parents and other taxpayers. For example, urban school districts often have poor tax bases and cannot readily raise funds for education. Suburban school systems have much less difficulty and can attract good teachers. There are continuing pressures for some suburban and urban school systems to merge or to share funds and resources, in the interests of both racial and financial equity. The health of inner city schools will be particularly important in encouraging minority youth to pursue science and engineering majors. The 44 largest urban school systems, represented by the Council of Great City Schools, enroll about 10 percent of the entire school population, but 33 percent of the Blacks and 27 percent of the Hispanics in public schools. These schools also enroll a disproportionately large number of students whose family incomes are below the poverty line.[44] Data from the National Assessment of Educational Progress (NAEP) assessments of science and mathematics achievement indicate that students in disadvantaged urban areas score an average of about 20 percent lower than the national average, while those in suburban areas score about 5 percent higher than the national average.[45]

[43]James S. Coleman and Thomas Hoffer, *Public and Private High Schools* (New York, NY: Basic Books, 1987) is a recent analysis of this proposition. Although this analysis does not specifically separate out mathematics and science education, the authors conclude that the ethos of the community surrounding a school, including taxpayers and parents, is more important in explaining the success of private schools than are particular actions that the school takes. Some argue that the political and bureaucratic milieux within which public schools operate harms them. See John E. Chubb and Terry M. Moe, "No School Is an Island: Politics, Markets, and Education," *The Brookings Review*, fall 1986, pp. 21-28. Other analysts have found that any advantage in the outputs of private school science experiences are balanced by the strong self-selectivity of private school students. See John R. Staver and Herbert J. Walberg, "An Analysis of Factors That Affect Public and Private School Science Achievement," *Journal of Research in Science Teaching*, vol. 23, No. 2, 1986, pp. 97-112.

[44]William Snider, "Urban Schools Have Turned Corner But Still Need Help, Report Says," *Education Week*, vol. 7, No. 2, Sept. 16, 1987, pp. 1, 20. Also see Bruce L. Wilson and Thomas B. Corcoran, *Places Where Children Succeed: A Profile of Outstanding Elementary Schools*, Report to U.S. Department of Education, Office of Educational Research and Improvement (Philadelphia, PA: Research for Better Schools, December 1987).

[45]U.S. Department of Education, op. cit., footnote 1, table 79. This statement is based on mathematics data for 1981-82 and science data for 1976-77. Due to cutbacks in the Department of Education's funding for the National Assessment of Educational Progress, a limited science assessment was conducted in 1981-82 with funding from the National Science Foundation. Data from that assessment were not tabulated by the geographic location of respondents, so cannot be used in this comparison. The 1986 assessment in mathematics shows continued gains by Black and Hispanic students in all three age categories (9, 13, and 17). See Dossey et al., op. cit., footnote 18.

Rural school systems face different problems in providing high-quality education, particularly in advanced mathematics and science, to geographically dispersed populations. While the days of the one-school school district are passing, rural districts still find it difficult to provide optional advanced mathematics and science courses and to attract the best teachers. These problems will grow worse in areas of rural America that continue to experience economic declines. Experiments are under way in some areas with distance learning technologies and regional science high schools.[46]

Standardized Achievement Testing

Students take many kinds of tests throughout their schooling to measure their learning and mastery of skills. Such tests are used to sort students among classes and tracks, to evaluate performance, to check for special abilities (see section in ch. 4 on programs for gifted and talented students) or learning disabilities, and to inform the college admissions process. Tests of basic competencies of high school graduates are growing in favor as part of the movement toward increased educational accountability. A 1985 survey found that 11 States required such tests of high school graduates, and 4 had plans to institute such tests.[47]

Many of these tests are of the familiar standardized multiple-choice type. Scores report students' progress both in absolute terms and relative to the performance of their peers. Such tests are inexpensive to administer; scoring is often done by computer.

But testing is controversial on several counts. It is said to deter many students from preparing for science and engineering careers. The tests are also said to convey racial, cultural, and gender biases against women and ethnic minorities that ostensibly lead to lower scores.[48] Some claim that testing has a pervasive harmful effect on the curriculum: teachers invite students to parrot back facts they have memorized, with the result that students' higher order thinking skills are not exercised.[49] A recent issue of the Newsletter of the National Education Association's Mastery in Learning Project put the issue most dramatically:

> Perhaps it has been the failure to understand intelligence—how it is nurtured or stunted, how it works, how it should be measured, even where it resides—that has done the most damage to the education of children.
>
> Because the workings and the vulnerability of the intellect have been so dimly understood by so many, teaching has often been rigidly fact-driven, heavily demanding of linguistic, logical, linear thinking skills, and often neglecting those aspects of learning that involve imagery, intuitiveness, manual and whole body skills, and feelings.
>
> Partly because of this failure of understanding, educational testing companies have continued to produce, states to require, and schools to administer, paper and pencil tests that, in purporting to measure students' achievement, have succeeded only in labeling and limiting it.[50]

The current practice of educational testing appears to constrict the pipeline for future scientists and engineers. It measures only a limited range of abilities and is often misused to deny students access to courses and encouragement that might tip the balance toward their becoming scientists and engineers. Alternative tests are being devised to address, in particular, knowledge of the processes of science, familiarity with experimental techniques, and higher order thinking skills, but the nagging problem will be resistance to their replication and large-scale adoption both for classroom use and in the college admissions process.[51]

[46]E. Robert Stephens, "Rural Problems Jeopardize Reform," *Education Week*, Oct. 7, 1987, pp. 25-26. A new federally sponsored project called ACCESS, in northwest Missouri, is designed to expand rural students' access to higher education in all subjects. Legislation has been introduced to set up similar programs in other States. See Robin Wilson, "U.S.-Backed Project in Missouri Aims to Help Rural Youths Overcome Farm Troubles and Continue Their Education," *The Chronicle of Higher Education*, Apr. 27, 1988, pp. A37-A38.

[47]U.S. Congress, Office of Technology Assessment, "State Educational Testing Practices," background paper, NTIS #PB88-155056, December 1987.

[48]Nevertheless, it is important to note that Asian students typically do better than other groups on the mathematics portion of these tests, indicating the difficulty of pinning down exactly what form any racial and cultural biases take.

[49]A comprehensive review is found in Norman Frederiksen, "The Real Test Bias: Influences on Teaching and Learning," *American Psychologist*, vol. 39, No. 3, March 1984, pp. 193-202.

[50]National Education Association Mastery in Learning Project, *Doubts and Certainties*, vol. II, No. 6, April 1988, p. 1.

[51]The Assessment of Performance Unit of Great Britain's Department of Education and Science, for example, has devised tests of students' skills in conducting experiments and in interpreting these

(continued on next page)

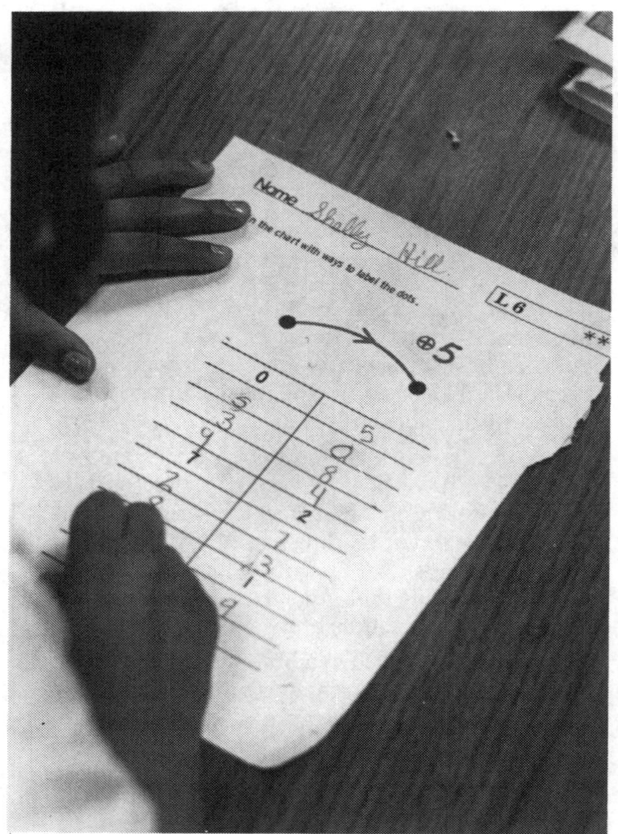

Photo credit: William Mills, Montgomery County Public Schools

Students are tested early and often in schools; this can help in evaluating their learning progress, but it also profoundly affects the curriculum.

The biggest controversies in testing, as it affects those who will major in science and engineering, have concerned the Scholastic Aptitude Test (SAT) and the American College Testing program, one of which almost all intending college students take prior to admission.[52] Because these tests are given great weight in admissions decisions by colleges and universities, any deficiencies in the preparation or administration of these tests could lead to "misassignment" of students among colleges and majors, or the failure of students to be admitted to college at all. Standardized tests have been important in college admissions for many decades, because they are economical to administer and measure students nationally against a common metric. But criticism of, in particular, the SAT by activist groups such as the Cambridge-based National Center for Fair and Open Testing (FairTest) and others in education is leading a small but increasing number of colleges and universities to drop their requirement for applicants to take the SAT.[53]

Females, Blacks, and Hispanics, on average, score lower than males and whites on the mathematics and verbal portions of these tests. In relation to disparities on the mathematics portion, some argue that this difference arises from the relatively poor preparation and limited number of mathematics and science courses taken by these groups. But others say that subtle biases in the tests' design and administration are a cause of some of the disparity.[54]

One important trend in testing is the increasing number of advanced placement programs being offered by schools. These programs give high school students college credit in a particular subject, based on the results of an examination which involves both multiple-choice and written responses. Many argue that this makes the advanced placement a better test, although it is more expensive than regular college admissions tests (costing over $50 per test). The number of such tests being taken is increasing about 13 percent annually, and about 20 percent of all secondary schools

(continued from previous page)
results. These tests have been replicated by the National Assessment of Educational Progress and in the advanced placement biology examination. Fran Blumberg et al., *A Pilot Study of Higher-Order Thinking Skills Assessment Techniques in Science and Mathematics* (Princeton, NJ: National Assessment of Educational Progress, November 1986). Also see Robert E. Yager, "Assess All Five Domains of Science," *The Science Teacher*, October 1987, pp. 33-37; and George E. Hein, "The Right Test for Hands-on Learning," *Science and Children*, October 1987, pp. 8-12.

[52]U.S. Congress, Office of Technology Assessment, *Educating Scientists and Engineers: Grade School to Grad School*, OTA-SET-377 (Washington, DC: U.S. Government Printing Office, June 1988), pp. 35-36. Also see Elizabeth Greene, "SAT Scores Fail to Help Admission Officers Make Better Decisions, Analysts Contend," *The Chronicle of Higher Education*, July 27, 1988, p. A20.

[53]David Owen, *None of the Above* (Boston, MA: Houghton Mifflin, 1985); Robert Rothman, "Admission Tests Misused, Says College Leader," *Education Week*, Dec. 9, 1987, p. 5. FairTest reports that at least 40 colleges now do not require either the Scholastic Aptitude Test or the American College Testing program for college admission. When colleges do not use these tests, they increase the weight that they place on other components of the admissions process, such as student transcripts, student essays, teachers' and counselors' recommendations, and in-person interviews. These components arguably allow candidates to present a much fuller impression of themselves as potential college students than do simple scores on standardized tests.

[54]See National Science Board, op. cit., foonote 25, p. 23, which sides with the "poor preparation" hypothesis.

and 10 percent of high school graduates now participate. About one-third of the examinations taken are in mathematics and science.[55]

Patterns of High School Course Offerings and Enrollments

Along with family encouragement and expectations, preparation in high school mathematics and science courses is vital to success in college-level science and engineering studies. Students' exposure to the traditional college-preparatory sequence of mathematics and science courses is restricted by both the course offerings of their schools and their willingness to take those courses. Minorities in particular often have less access to advanced mathematics and science courses, because school districts with high minority enrollments often cannot afford to offer many such courses. Offerings in rural and urban schools are generally more limited than those in suburban schools.

Even when advanced courses are offered, students who could benefit from them often fail to take them. The point at which students are first allowed to decide which mathematics or science courses they are to take is widely believed to be an important fork in the educational pipeline for future scientists and engineers. Once students fail to pursue the normal preparatory sequence of courses, it becomes hard for them to catch up.

Research indicates that there is a positive correlation between the number of advanced high school mathematics and science courses taken and two educational outcomes: achievement test scores and students' intentions to major in science and engineering.[56] Correlations between mathematics or science course-taking and achievement test scores have been found in analyses of data from both the NAEP mathematics assessment of 1982 (for mathematics courses) and the 1980 High School and Beyond (HS&B) survey of the sophomore cohort (for both mathematics and science courses).[57] These findings were sustained even after statistical allowance was made for student's race, ethnicity, socioeconomic status, and earlier test scores. A strong correlation between high school mathematics course-taking and the major choices of college students was found in an OTA analysis of the 1980 HS&B cohort, even when many other factors were statistically controlled.[58]

Mathematics course-taking, presumably due to its sequential nature, appears to be more important for success in later science and engineering study than science course-taking. The College Board recently noted in *Academic Preparation for Science*, a handbook that advises high school teachers about what colleges would like them to teach, that knowledge of scientific skills and fundamental concepts will be more important to students than the number of high school science courses they have completed.[59]

It is difficult to be sure that the number of mathematics and science courses taken is a principal influence in the decision to major in science or engineering.[60] However, such courses do keep students in the pipeline.

Course Offerings and Enrollments

Students cannot take courses their schools do not offer. Only a few schools offer the complete

[55]Jay Mathews, "Tests Help 'Ordinary' Schools Leap Ahead," *Washington Post*, May 14, 1987. In 1985-86, 7,201 schools participated out of 30,000 secondary schools nationally. Garfield High School in Los Angeles, featured in the recent movie "Stand and Deliver," has become one of the top 10 schools in the Nation for the number of students who take and pass the advanced placement calculus test. Garfield is located in a poor and predominantly Latino neighborhood.

[56]The correlation between outcomes such as these and the number of mathematics courses tends to be stronger than that with the number of science courses.

[57]Josephine D. Davis, *The Effect of Mathematics Course Enrollment on Racial/Ethnic Differences in Secondary School Mathematics Achievement* (Princeton, NJ: Educational Testing Service, January 1986); and Lyle V. Jones et al., "Mathematics and Science Test Scores as Related to Courses Taken in High School and Other Factors," *Journal of Educational Measurement*, vol. 23, No. 3, fall 1986, pp. 197-208.

[58]Valerie E. Lee, "Identifying Potential Scientists and Engineers: An Analysis of the High School-College Transition," OTA contractor report, 1987.

[59]The College Board concludes that the amount of high school science course-taking makes relatively little difference to students' subsequent college performance in science. OTA is skeptical of this conclusion. See below and College Entrance Examination Board, *Academic Preparation in Science: Teaching for Transition From High School to College* (New York, NY: 1986), pp. 14-16. See also Robert E. Yager, "What Kind of School Science Leads to College Success?" *The Science Teacher*, December 1986, pp. 21-25.

[60]College Entrance Examination Board, op. cit., footnote 59, pp. 14-16.

range of college-preparatory mathematics and science courses, a deficiency that has persisted for many years.[61] Data on course offerings and enrollments are plagued with inconsistencies. For example, courses with the same titles may have different content while those with near-identical content may have different titles. Inconsistencies in data make the task of comparing schools, States, and years very difficult.

The most recent data on course offerings comes from the 1985-86 National Survey of Science and Mathematics Education, sponsored by the National Science Foundation.[62] This survey used the same course classification system as a 1977 survey, permitting comparisons over time. Over 90 percent of high schools offer at least algebra I, algebra II, and geometry, but advanced course offerings are more limited. Only about 31 percent of schools offer a full calculus course (although some senior-year mathematics courses may include an introduction to calculus), and 18 percent offer a course leading to the advanced placement examination in calculus. In science, over 90 percent of high schools offer at least 1 year of biology and chemistry, and 80 percent offer 1 year of physics.

Since 1977, mathematics course offerings have increased somewhat, though the proportion of schools offering calculus has remained constant. Science course offerings have increased slightly. In general, schools offer only one section of these college-preparatory courses. It is not clear whether this outcome restricts or reflects demand for such courses. For example, 23 percent of U.S. high schools offer only one section of biology and 52 percent offer only one section of physics.[63]

Data on course enrollments in mathematics and science indicate that the proportion of high school graduates that have taken college-preparatory mathematics and science courses is very small (again see figure 2-3). While 77 and 61 percent of students took algebra I and geometry, respectively, only 20 percent took trigonometry and only 6 percent took calculus. In science, 90 percent took biology, but only 45 percent took chemistry I and 20 percent took physics. All of these proportions (except for calculus) represent increases from 1984.[64]

Another analysis of the same database, which used a somewhat different course classification, suggests that, even where courses are offered, enrollments are low. About 80 percent of the students to whom the course is available enroll in algebra I, 48 percent in geometry, and about 20 percent in trigonometry. Similarly, in science, while almost all students to whom it was offered took biology, about one-third of the students with the chance to take chemistry did so as did only 10 percent of the students offered physics.[65]

More recent data from the 1985-86 NAEP in mathematics provide a "snapshot" view of enrollments in mathematics classes (see table 2-3, but note that the classification of courses used here differs from that used in other tables). These data suggest that advanced course-taking in mathematics remains at a small proportion of 17-year-olds, although there were some very small increases between 1982 and 1986. Because these data were taken from 17-year-olds, who have the senior year of high school to go before graduation, they do not provide a complete picture of high school course-taking.

These findings send a clear message: offerings of pipeline mathematics and science courses are constrained. More importantly, even when they are offered, only tiny numbers of students take them.

[61]Some advanced courses are offered to high school students by community colleges, but there are no national data on this phenomenon.

[62]Weiss, op. cit., footnote 15. No separate data are available on offerings and enrollments in laboratory courses; data on the amount of time different mathematics and science classes spend on laboratory work is included in ch. 3.

[63]These data are confirmed by a National Science Teachers Association survey. See Bill G. Aldridge, "What's Being Taught and Who's Teaching It," *This Year in School Science 1986: The Science Curriculum*, Audrey B. Champagne and Leslie E. Hornig (eds.) (Washington, DC: American Association for the Advancement of Science, 1987), ch. 12.

[64]Westat, op. cit., footnote 21. Data from 1982 are presented in the U.S. Department of Education, National Center for Education Statistics, "Science and Mathematics Education in American High Schools: Results From the High School and Beyond Survey," NCES 84-211b, Bulletin, May 1984, tables A-3, A-4, A-5. In general, Asian students are two to four times as likely to take advanced biology, chemistry, and physics courses than other minority students.

[65]Evaluation Technologies, Inc., op. cit., footnote 9.

Table 2-3.—Trends in Mathematics Course-Taking, 1982-86

Course	Percentage of 17-year-olds by the highest level of mathematics course they have taken						
	Year	Total	Males	Females	Black	Hispanic	White
Pre-algebra	1982	24	25	24	34	37	32
	1986	19	19	19	31	25	17
Algebra I	1982	16	16	17	20	21	15
	1986	18	17	18	18	24	17
Geometry	1982	14	13	15	10	12	15
	1986	17	15	18	16	16	17
Algebra II	1982	39	39	39	29	24	41
	1986	40	39	40	31	28	42
Pre-calculus or calculus	1982	5	6	5	4	3	5
	1986	7	8	5	3	6	7

SOURCE: John A. Dossey et al., *The Mathematics Report Card: Are We Measuring Up? Trends and Achievement Based on the 1986 National Assessment* (Lawrence Township/Princeton, NJ: Educational Testing Service, Inc., June 1988), table 8.2.

Females and Minorities Lag in Course-Taking

Females, Blacks, and Hispanics, according to the HS&B survey, fall behind their white male peers in enrollments in high school advanced mathematics and science courses. This finding is confirmed in data collected for the 1985-86 NAEP mathematics and science assessments.[66] Tables 2-4 and 2-5 show that, as one follows the normal

[66] For mathematics data, see Dossey et al., op. cit., footnote 18, pp. 116-117. Science data are due to be published in fall 1988.

Table 2-4.—Percentage of 1982 High School Graduates Who Went on to Next "Pipeline" Mathematics Course After Completing the Previous Course

	Percentage that took geometry after passing algebra	Percentage that took algebra II after passing geometry	Percentage that took trigonometry after passing algebra II	Percentage that took calculus after passing trigonometry
Sex:				
Males	67	55	43	31
Females	63	52	34	30
Of those earning As or Bs on previous course, by sex:				
Males	82	62	55	47
Females	74	61	45	41
Race/ethnicity:				
Hispanics	50	47	28	28
Black	57	55	29	29
White	67	54	40	30
Of those earning As or Bs on previous course, by race/ethnicity:				
Hispanic	64	56	45	48
Black	74	62	48	31
White	79	62	50	43
Urbanicity of school:				
Urban high school	65	55	36	25
Suburban high school	69	52	40	33
Rural high school	58	55	39	26
Regional differences:				
New England	76	76	32	50
Mid-Atlantic	64	54	48	40
West North Central	66	45	32	9
West South Central	53	62	32	19
Curricular track:				
General	52	41	27	9
Academic	82	61	45	36
Vocational	43	35	19	11

NOTE: The source from which this tabulation is derived does not include the total numbers of students in these samples. Also the data (as originally reported) do not indicate the actual *order* in which the courses were taken, only that the students had taken these courses before graduating from high school. To this extent, the tabulation forces an artificial formalism on the order of course-taking.

SOURCE: C. Dennis Carroll, *Mathematics Course Taking by 1980 High School Sophomores Who Graduated in 1982* (Washington, DC: U.S. Department of Education, National Center for Education Statistics, April 1984).

Table 2-5.—Percentage of 1982 High School Graduates Who Went on to Next "Pipeline" Science Course After Completing the Previous Course

	Percentage that took biology after passing general science	Percentage that took chemistry after passing biology	Percentage that took physics after passing chemistry
Sex:			
Males	72	39	47
Females	75	37	31
Of those earning As or Bs on previous course, by sex:			
Males	79	59	61
Females	79	50	40
Race/ethnicity:			
Hispanics	71	21	33
Black	74	28	27
White	74	42	40
Of those earning As or Bs on previous course, by race/ethnicity:			
Hispanic	75	32	43
Black	78	43	41
White	79	57	51
Urbanicity of school:			
Urban high school	72	33	39
Suburban high school	73	41	40
Rural high school	75	36	37
Regional differences:			
New England	76	47	44
Mid-Atlantic	74	49	44
West South Central	83	29	20
Mountain	74	28	35
Curricular track:			
General	71	21	23
Academic	83	59	44
Vocational	64	15	22

NOTE: The source from which this tabulation is derived does not include the total numbers of students in these samples. Also the data (as originally reported) do not indicate the actual *order* in which the courses were taken, only that the students had taken these courses before graduating from high school. To this extent, the tabulation forces an artificial formalism on the order of course-taking.

SOURCE: Jeffrey A. Owings, *Science Course Taking by 1980 High School Sophomores Who Graduated in 1982* (Washington, DC: U.S. Department of Education, National Center for Education Statistics, April 1984).

sequence of mathematics and science courses designed as preparation for college-level study in science and engineering, there is constant attrition in all categories of students. The attrition of females, Blacks, and Hispanics is much greater than that of white males. Black and Hispanic 17-year-olds instead are more likely to report that their highest mathematics course was pre-algebra than white students.

For example, while 5.6 percent of all high school graduates in 1982 took calculus, only 2 percent of Blacks and 2.4 percent of Hispanics did. The situation showed little change by 1986, according to NAEP data, although Hispanic students had doubled their participation in pre-calculus or calculus classes in that time. The gender difference, however, is not so pronounced: 5 percent of females and 6.1 percent of males take calculus. Another way of examining these data is in terms of the proportion of students—by sex, race, and ethnicity—that go on to take a higher mathematics or science course after successfully completing the last one. Tabulations of these transition percentages, based on a Department of Education analysis of the HS&B survey, appear in tables 2-6 and 2-7.

The data clearly show that females and minorities drop out of the normal sequence of courses. In mathematics, females drop out after taking algebra II and fail to take trigonometry; they also forgo physics after taking chemistry. Blacks and Hispanics similarly fall out after trigonometry, although fewer of them move from algebra to geometry than do whites. These disparities are stronger among the "high-talent" groups of those who earned As and Bs on the previous courses.

Table 2-6.—Percentage of 1982 High School Graduates Who Have Taken College Preparatory Mathematics Courses by Sex, and Race/Ethnicity

Subject	All	Males	Females	Asian	Black	Hispanic	White
Algebra I	63	61	65	65	53	54	66
Algebra II	31	31	31	44	22	19	34
Geometry	48	47	49	68	33	28	53
Trigonometry	7	9	6	16	4	5	8
Other advanced mathematics	13	14	13	30	5	7	15
Calculus	6	6	5	15	2	2	6

SOURCE: U.S. Department of Education, National Center for Education Statistics, "Science and Mathematics Education in American High Schools: Results From the High School and Beyond Survey," NCES 84-211b, Bulletin, May 1984, table A-5.

Table 2-7.—Percentage of 1982 High School Graduates Who Have Taken College Preparatory Science Courses by Sex, and Race/Ethnicity

Subject	All	Males	Females	Asian	Black	Hispanic	White
General science	30	30	30	24	33	34	29
Basic biology	74	73	76	78	74	69	75
Advanced biology	8	7	9	13	6	5	9
Chemistry I	24	25	24	41	19	13	27
Advanced chemistry	4	5	3	8	2	2	4
Geology	14	15	13	9	11	12	15
Physics I	11	15	8	27	6	5	13
Advanced physics	1	2	1	5	1	1	2
Unified science	28	30	26	17	34	21	27

SOURCE: U.S. Department of Education, National Center for Education Statistics, "Science and Mathematics Education in American High Schools: Results From the High School and Beyond Survey," NCES 84-211b, Bulletin, May 1984, tables A-4 and A-5.

High-talent females drop out after algebra I, algebra II, and trigonometry. Very few high-talent Blacks take calculus, compared to whites, possibly indicating the paucity of calculus offerings in schools with high minority populations. Hispanics' persistence in science and mathematics course-taking is low throughout.

The data underscore the advantage students gain from attending suburban high schools rather than urban or rural ones. They also show a significant geographic disparity in persistence in mathematics and science course-taking. Persistence rates in the New England and Mid-Atlantic regions are some 50 or more percent higher than in West North Central, West South Central, and Mountain regions. (See figures 2-6 and 2-7.)

These findings are supported by an analysis of data from the 1982 NAEP mathematics assessment, which found that, regardless of curricular track, racial composition of the school attended, grade or achievement level, Black and Hispanic students lagged in enrollments in advanced mathematics classes compared with their white peers.[67]

Black enrollments in high school science courses vary significantly according to geographic region, parental expectations (especially those of mothers), and high achievement in other subjects such as English.[68] Black students in the Mid-Atlantic regions were more likely to take science courses, while those in the Pacific and the East South Central regions were least likely to take science courses.

[67]Davis, op. cit., footnote, 57, p. 74.

[68]Ellen O. Goggins and Joy S. Lindbeck, "High School Science Enrollment of Black Students," *Journal of Research in Science Teaching*, vol. 23, No. 3, 1986, pp. 251-261.

TRACKING AND ABILITY GROUPING

Ability grouping is practiced nearly universally in American schools. Guidance counselors and teachers sort students by ability as early as the third grade, using standardized tests and individual judgment. Students who display the conventional attributes of the potential scientist or engineer are encouraged to pursue the mathematics and science courses that will prepare them for

Figure 2-6.—Percentage of 1982 High School Graduates Who Took Calculus

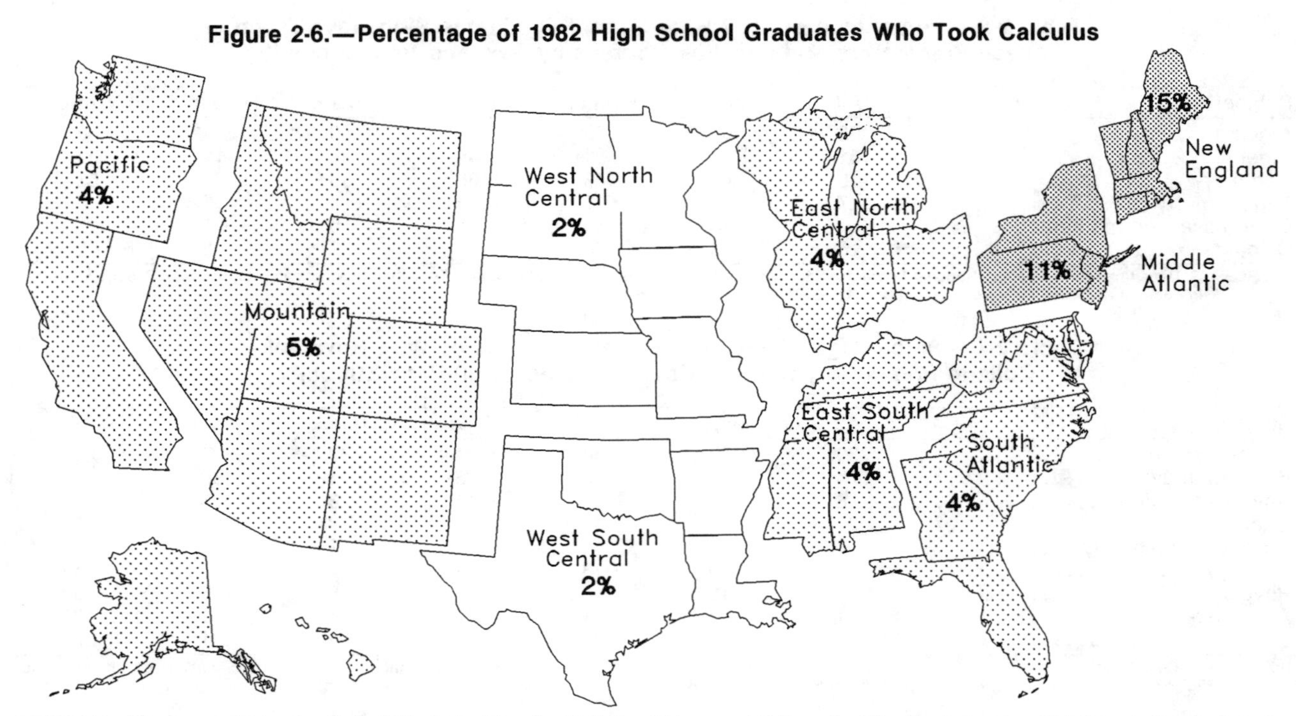

SOURCE: U.S. Department of Education, National Center for Education Statistics, "Science and Mathematics Education in American High Schools: Results From the High School and Beyond Study," NCES 84-211b, Bulletin, May 1984, table A-4.

Figure 2-7.—Percentage of 1982 High School Graduates Who Took Physics

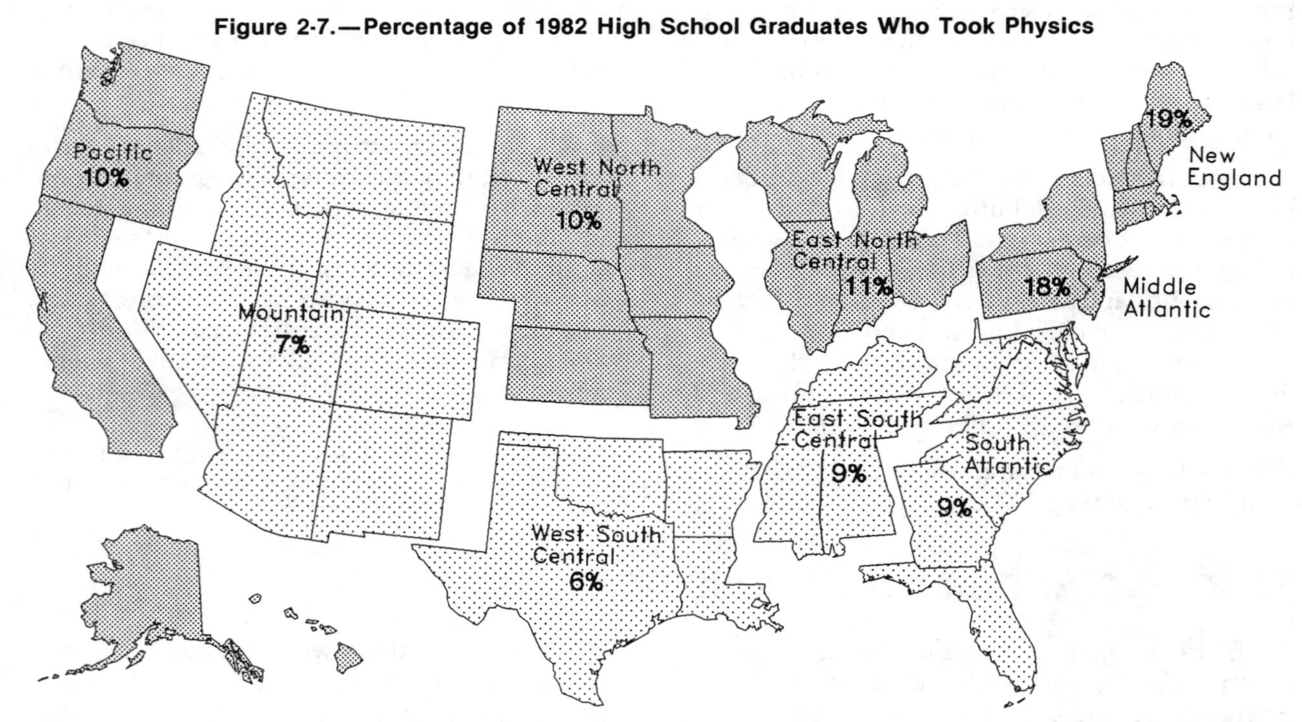

SOURCE: U.S. Department of Education, National Center for Education Statistics, "Science and Mathematics Education in American High Schools: Results From the High School and Beyond Study," NCES 84-211b, Bulletin, May 1984, table A-3.

these careers. Proponents claim that students' interests are reinforced by exposure to those of their similarly enthusiastic peers. Others are directed toward other courses and careers. What we have here is a double-edged sword.

The Double-Edged Sword

Tracking or grouping is intended to help prevent the quick learner from trampling over the slow learner and the slow from delaying the progress of the quick. In short, tracking is supposed to give all students a fair chance (and help teachers maintain control). Comparisons and labels, however, are inevitable. Tracking is widely believed to both harm and help students' self-esteem, progress, achievement, socialization, and educational and vocational destinations. Because its effects are varied and not readily measurable, it has been a very difficult issue for educational researchers to study.

Many students' career options are narrowed by this sorting. Those who fail to display the signs of early promise, and those whose home life or idiosyncracies place academic or social obstacles in their paths, may find themselves shunted aside from the mathematics and science preparation that makes possible further study of science or engineering. Many of these students have the ability and the desire to pursue these careers. About one-quarter of those who go on to major in science or engineering were in a nonacademic track in high school. Generally, their relatively poor mathematics and science preparation makes it difficult for them to keep up in college, and they are at risk of attrition. Thus, ability grouping, if applied too clumsily or rigidly, may lead to the waste of talent.

Minority leaders, both within and outside the education community, have complained that tracking perpetuates racism. The evidence is that the practice is a structural impediment to students' progress to advanced study in science and engineering. Nevertheless, inconsistencies in definitions between surveys that include information on tracking often yield misleading comparisons. In particular, it is not possible to state with any certainty whether enrollments in the academic or general tracks, either in total or by race or gender, are declining through time. Figure 2-8, taken from High School and Beyond data, shows the proportion of students, by race and gender, that were enrolled in each high school curriculum track in 1982, but these data are not necessarily consistent with other surveys.[69] It is clear, however, that students are enrolling in courses in much more varied patterns than they used to: they frequently mix courses designed for the general, vocational, or academic track. To this extent, the stranglehold of tracking is loosening.

The principal objection to tracking or ability grouping is that it can become a self-fulfilling prophecy, changing the behavior of students, students' peers, teachers, and parents toward members of a particular group. For slow-tracked students the technique stifles aspiration by rein-

[69]Similar data are reported in Davis, op. cit., footnote 57, p. 23, based on the National Assessment of Educational Progress (NAEP) 1981-82. More recent data from the 1985-86 NAEP mathematics assessment put the proportion of 17-year-olds enrolled in the academic track at 52 percent, in the general track at 38 percent, and in the vocational track at only 10 percent. Dossey et al., op. cit., footnote 18, p. 119.

Figure 2-8.—Track Placement by Race/Ethnicity and Sex, High School Graduates of 1982

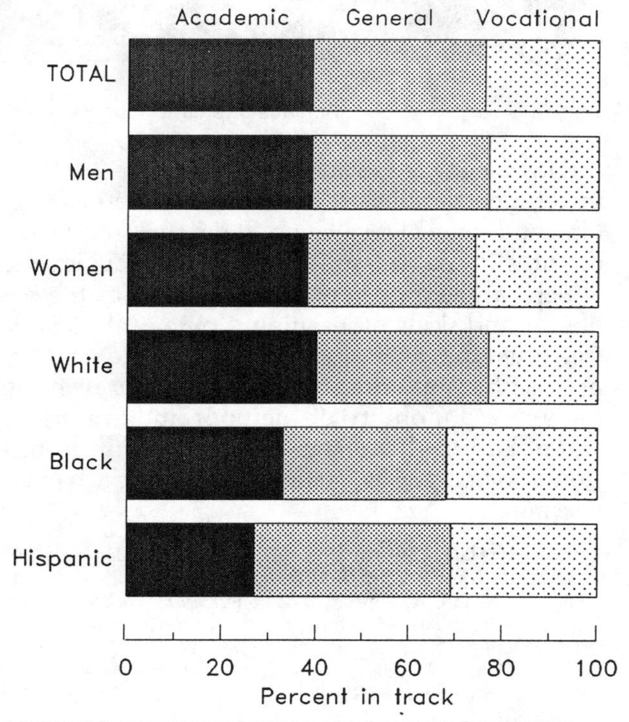

SOURCE: U.S. Department of Education, High School and Beyond survey.

Photo credit: William Mills, Montgomery County Public Schools

Science-intensive schools often have sophisticated, costly equipment.

forcing feelings of low status and, worse, by feeding students' beliefs that they have been left behind and can never catch up. It deprives this group of the stimulation provided by high achievers, which can help promote development of the behavior and skills of learning. Some writers have suggested that grouping and tracking is a primary means of maintaining the status quo, preventing the upward mobility of the poor and minorities and excluding them from preparation for professional occupations such as science and engineering.[70]

[70]Bruce R. Hare, "Structural Inequality and the Endangered Status of Black Youth," *Journal of Negro Education*, vol. 56, No. 1, 1987, pp. 100-110; Joel Spring, *The American High School 1942-1985: Varieties of Historical Interpretation of the Foundations and Development of American Education* (New York, NY: Longman, 1986); and William Snider, "Study Examines Forces Affecting Racial Tracking," *Education Week*, Nov. 11, 1987, pp. 1, 20.

Effects of Grouping Practices

For all the controversy, research on the effects of grouping is far from conclusive, even when it uses simple output measures such as achievement scores. This inconclusiveness might be taken as evidence that other important factors, such as students' socioeconomic status and teaching quality, are more important in predicting educational outcomes than the existence of tracking.

Research on the effect of ability-based grouping practices in elementary schools has found that grouping makes little or no difference to the most able students, but does have a considerable retarding effect on the less able students.[71] A 1968

[71]Robert E. Slavin, "Ability Grouping and Student Achievement in Elementary Schools: A Best-Evidence Synthesis," prepared for

study of tracking in secondary schools by the National Education Association found that for each study that showed a gain in achievement scores across the ability spectrum another study showed a net loss. The exception was the lowest ability level, which had uniformly slightly more losses than gains.[72]

To examine the effects of tracking on students intending majors in science and engineering, OTA used the High School and Beyond database.[73] In the survey's random sample of about 12,000 high school sophomores in 1980, 25 percent of those students planning science and engineering majors by their senior year and scoring above average on the HS&B achievement test had been enrolled in the general and vocational tracks. Compared with their academically tracked peers, this group was of lower socioeconomic status and had a slightly lower average achievement test score. By the end of high school, they had taken about one less mathematics course and their overall grade point average was about one-quarter of a point lower. Their average SAT score was about 68 points lower. They were less likely to go to college and more likely to enroll in a junior college than members of the academically tracked group. Table 2-8 displays some characteristics of the two groups.[74]

Table 2-8.—Science-Intending Students Among High School Graduates of 1982, by Track

Characteristics	Group from nonacademic tracks N = 428	Group from academic track N = 1,147
Demographic characteristics:		
Percent Black	3	5
Percent Hispanic	6	5
Percent female	40	41
High school experiences:		
Score on HS&B Achievement Test[a]	55.9	59.2
Number of advanced mathematics courses taken	2.0	3.1
Average high school grade point average	2.8	3.0
Score on mathematics portion of the SAT or ACT[b]	457	525
College experiences:		
Percentage who had enrolled in college by February 1984	67	89
Percentage in 2-year colleges	47	24
College grade point average	2.8	2.9

KEY: HS&B = High School and Beyond survey.
SAT = Scholastic Aptitude Test.
ACT = American College Testing program.

[a]On HS&B Achievement Test, mean score=50, standard deviation=10.
[b]Scores are normalized to those for the SAT, with a range of 0 to 800.

SOURCE: Valerie Lee, "Identifying Potential Scientists and Engineers: An Analysis of the High School-College Transition," OTA contractor report, September 1987; based on the High School and Beyond survey.

Other data indicate that academically tracked 17-year-olds are more than twice as likely than those in other tracks to survive to algebra II in the normal sequence of high school mathematics courses, and about five times as likely to survive to pre-calculus or calculus.[75] Tracking does have some positive effects on the academically tracked science-intending stream, however, for it generally ensures their continuing participation and preparation in the science and engineering pipeline, by increasing the probability that they will take pipeline mathematics and science courses.

For those who run afoul of the system—by reason of race, class, attitude, or bias—access to high-quality, academic mathematics and science

the U.S. Department of Education, Office of Educational Research and Improvement, Grant No. OERI-G-86-0006, June 1986.

[72]Robert E. Fullilove, "Images of Science: Factors Affecting the Choice of Science as a Career," OTA contractor report, 1987. The National Education Association study is quoted in James E. Rosenbaum, "Social Implications of Educational Grouping," *Review of Research in Education*, David Berliner (ed.), vol. 8 (Itasca, IL: F.E. Peacock Publishers, 1980), pp. 361-401. For other research, see Glenna Colclough and E.M. Beck, "The American Educational Structure and the Reproduction of Social Class," *Sociological Inquiry*, vol. 56, No. 4, fall 1986, pp. 456-476; Beth E. Vanfossen et al., "Curriculum Tracking and Status Maintenance," *Sociology of Education*, vol. 60, April 1987, pp. 104-122; and Gerald W. Bracey, "The Social Impact of Ability Grouping," *Phi Delta Kappan*, May 1987, pp. 701-702.

[73]Lee, op. cit., footnote 58.

[74]A regression analysis indicated that, for these students, track placement was a stronger predictor than class, race and ethnicity, or gender of the number of academic mathematics courses that students took in high school. From a national perspective on the production of scientists and engineers, this finding attests to the centrality of students' preparation in high school mathematics. An important new national longitudinal survey being conducted by Jon Miller at Northern Illinois University, funded by the National Science Foundation, should disentangle many of the influences of early mathematics and science learning. The study is following a cohort from grade eight onwards, and is surveying family, social, and school variables that might affect science and mathematics learning, attitudes, and behaviors.

[75]Dossey et al., op. cit., footnote 18, table 8.3.

courses is lost and their expectations are dulled. Nevertheless, the academic track is not the right place for all students. A corollary problem is the early age at which tracking occurs, putting many students at a considerable disadvantage when they enter middle and high school. The need is for systems to practice tracking efficiently but flexibly.

Science Education and Track-Busting

In the context of science and mathematics education, the "efficiency" and "flexibility" of tracking may be incompatible with the notion—and today the more commonly heard prescription—of "science for all."

> Unless students very early acquire both a basic conceptual science vocabulary and a zest for learning and problem solving, they are extremely unlikely to take science courses—or to succeed if they do. . . . [Needed, then, is] a baseline science curriculum that will provide all students with a consistent and coherent overview and an integrated body of knowledge during the elementary and high school years.[76]

What may appear to be "special pleading" for science and mathematics might then also be seen as one rationale for "track-busting." Put another way, a change in expectations of students' capability will have to precede changes in both teaching and learning.

[76] George W. Tressel, "Priestley Medal Address" (letter), *Chemical & Engineering News*, Sept. 19, 1988, pp. 3, 39. Also see George W. Tressel, "A Strategy for Improving Science Education," presented to the American Educational Research Association, Apr. 8, 1988.

Chapter 3
Teachers and Teaching

Photo credit: William Mills, Montgomery County Public Schools

CONTENTS

Boxes

Figures

Tables

Chapter 3
Teachers and Teaching

I'll make a deal with you. I'll teach you math, and that's your language. With that you're going to make it. You're going to go to college and sit in the first row, not in the back, because you're going to know more than anybody.

Jaime Escalante, 1988

America's schools will shoulder important new responsibilities in the years to come. Well-educated workers of all kinds are looked on increasingly as economic resources.[1] Schools, parents, communities, and government at all levels are expected to educate a population that will grow more ethnically diverse in an economy that is increasingly reliant on science and technology. International competition will be invoked as a spur to excellence. The need for full participation by minorities and females will become a chronic national concern. Nowhere will these pressures be felt more strongly than in the education of scientists and engineers. The pressures, in short, will fall squarely on mathematics and science teachers.

It is a burden the teaching profession, together with school districts and teacher education institutions, is ill-equipped to meet. Fears of shortages of mathematics and science teachers now and in the future abound, and there is great concern about the poor quality of teacher training and in-service programs. The quality of teaching, in the long run, depends on the effectiveness of teachers, the adequacy of their numbers, and the extent to which they are supported by principals, curriculum specialists, technology and materials, and the wider community. Teachers of mathematics and science need to be educated to high professional standards and, like members of other professions, they need to update their skills periodically.

At the same time, research on the teaching of mathematics and science suggests that some techniques not widely used in American schools can improve achievement, transmit a more realistic picture of the enterprise of science and mathematics, and broaden participation in science and engineering by females and minorities. These techniques have been adopted slowly. Practical experiments and class discussion, for example, are slighted by many teachers in favor of lectures, book work, and "teaching to the test." Small group learning, in which students cooperate to accomplish tasks, is also rare, although a few States are making room for it in their educational prescriptions. Teachers themselves seldom have opportunities to exchange information with their colleagues in other schools. Increasing such opportunities—for teachers and students alike—could have significant effects on, among other things, the size and quality of the national science and engineering talent pool.

[1]See, for example, National Commission on Excellence in Education, *A Nation at Risk* (Washington, DC: U.S. Government Printing Office, April 1983).

THE TEACHER WORK FORCE

Without a teacher to explain, respond, and excite students' interest, formal education is dull and limited. Scientists and engineers tell many stories about their inspiring teachers.[2] Yet the effect

[2]A decade-old series of autobiographies sponsored by the Alfred P. Sloan Foundation, including books by Freeman Dyson, Peter Medawar, Lewis Thomas, S.E. Luria, and Luis W. Alvarez, have been resoundingly successful at capturing the ". . . perceptions of the individual who did the science—of how it was done," and are designed to be ". . . important for the next generation of scientists in high school and college." See John Walsh, "Giving the Muse a Helping Hand," *Science*, vol. 240, May 20, 1988, pp. 978-979. The latest in the Sloan series is by 1986 Nobel laureate Rita Levi-Montalcini, *In Praise of Imperfection: My Life and Work* (New York,

(continued on next page)

that a good mathematics and science teacher has on a student's propensity to major in science and engineering cannot easily be evaluated quantitatively.

There are two major, and related, challenges that affect mathematics and science education: the first is the potential for a shortage of mathematics and science teachers, and the second is the need to improve the quality of teaching. Some fear that States and school districts will simply lower certification and hiring criteria standards in the face of possible shortages. Shortages are likely to cause problems in certain States and school districts, especially in the supply of minority mathematics and science teachers. But improving the quality of mathematics and science teaching is as important as addressing shortages.

Science and mathematics teachers are part of the entire teaching work force. In many ways, there are few differences between mathematics and science teachers and teachers of other subjects. Each are covered by the same labor contracts, belong to the same teacher unions, share the same working conditions, and are normally paid the same salaries.[3] Similarly, mathematics and science teachers share in the low esteem with which many Americans hold teaching and public education in general.[4]

In mathematics and science teaching, there are important differences between teacher preparation and assignments in elementary schools and secondary schools. Elementary teachers teach many unrelated subjects, while secondary teachers concentrate on particular subjects, such as mathematics or science (although many do both, or teach several different science fields). Accordingly, most elementary teachers are not specialists in any subject. They normally hold baccalaureate degrees in education and have had relatively little science and mathematics coursework (if any) at college. Most secondary teachers, however, have taken many mathematics and science courses in college; some have an undergraduate degree in these disciplines.[5]

The Possibility of Mathematics and Science Teacher Shortages

Many observers are worried about possible future shortages of teachers, and, reportedly, in some geographic areas it is already difficult to hire adequate numbers of mathematics and science teachers.[6] It is widely believed that shortages of

(continued from previous page)

NY: Basic Books, 1988). Also see Daryl E. Chubin et al., "Science and Society," *Issues in Science & Technology*, vol. 4, summer 1988, pp. 104-105.

[3]An ongoing controversy related to the entire teacher work force is the role of unions. Some people think that teacher unions, through their sometimes stubborn resistance to change, are the cause of many problems in education. These problems include the difficulty in firing poor teachers and in staffing "difficult" schools, the devotion to the "seniority" principle (rather than teacher's merits) shown in allocating salary increases, and the potential barrier to meaningful reforms erected by the granting of "tenure" to teachers. Others think that teacher unions can be of great help in providing a single point of negotiation for many aspects of teachers' working lives and conditions, forging teachers into a profession based on common, self-specified norms and goals of conduct, and encouraging teachers to become more reflective of their tasks. There are two main teacher unions, the American Federation of Teachers and the National Education Association. Their leaders are visible in the national debate on reforming American education, often calling for greater public spending on education, and their positions have frequently been at odds with those of the U.S. Secretary of Education. There is no indication that the form and extent of union activity in mathematics and science teaching is any different from that for teaching as a whole (although there are special professional associations of such teachers, such as the National Science Teachers Association and National Council of Teachers of Mathematics). The positive and negative impacts of teacher unions are not considered further in this report.

[4]For example, surveys show that the percentage of Americans that would like their children to become public school teachers has fallen from 75 percent in 1969 to 45 percent in 1983. In a similar survey in 1981, Americans ranked clergymen, medical doctors, judges, bankers, lawyers, and business executives as being in professions with higher prestige and status than public school teaching. Only local political officeholders, realtors, funeral directors, and advertising practitioners were ranked lower. Stanley M. Elam (ed.), *The Phi Delta Kappa Gallup Polls of Attitudes Toward Education 1969-1984: A Topical Summary* (Bloomington, IN: Phi Delta Kappa, 1984).

[5]Most new teachers were education majors in college. Many, however, were single subject (such as physics) majors directly inducted into the teaching force or are taking supplementary education courses. The utility of the education major is under serious reconsideration at the moment and several groups have proposed a wide-ranging overhaul of teacher education. This is discussed later under "Preservice Education."

[6]See National Science Board, *Science and Engineering Indicators —1987* (Washington, DC: U.S. Government Printing Office, 1987), pp. 27-32; and Linda Darling-Hammond, *Beyond the Commission Reports: The Coming Crisis in Teaching*, RAND/4-3177-RC (Washington, DC: Rand Corp., July 1984). Henry M. Levin, Institute for Research on Educational Finance and Governance, School of Education, Stanford University, "Solving the Shortage of Mathematics and Science Teachers," January 1985, finds that shortages, in some form, have existed for 40 years, primarily because of the low salaries offered to mathematics and science teachers.

teachers of all kinds are imminent due to an increase in the number of teachers approaching retirement and a decrease in the number of college freshmen planning to become teachers during the last decade. In the aggregate, these trends affect the size of the teacher work force. But it is events in the middle stages of teachers' careers as well that predict future supply and demand. For example, many fully qualified teachers leave the profession (perhaps to start families), and may be lured to return to schools in due course.

To estimate whether there will be a shortage, and what its effects might be, it is necessary to have data on the future work plans of the existing teaching work force, the rates of entry to and exit from it, the extent to which these rates change in response to market signals, and what measures might reduce the effect of any shortage. A conceptual model of entry to and exit from the teacher work force is depicted in figure 3-1.

Current estimates of the rates of entry to and exit from the teaching work force are very poor and often inconsistent.[7] It is not possible, therefore, to determine with any certainty whether

[7]Lynn Olson and Blake Rodman, "Is There a Teacher Shortage? It's Anyone's Guess," *Education Week*, June 24, 1987, pp. 1, 14-16; and Blake Rodman, "Teacher Shortage Is Unlikely, Labor Bureau Report Claims," *Education Week*, Jan. 14, 1987, p. 7. Data, much of it conflicting, is collected and reported by the National Education Association, the U.S. Bureau of Labor Statistics, and the Department of Education. The inadequacy of databases on teachers is also revealed through the absence of reliable estimates of the number of uncertified teachers teaching or the number teaching outside their field of certification. The Center for Education Statistics is conducting a new survey, the results of which should be available in early 1989. Simultaneously, the National Academy of Sciences is examining future research needs on this issue, while the Council of Chief State School Officers is also trying to assemble disparate State data.

Figure 3-1.—Career Paths of 100 Newly Qualified Teachers, About 1 Year After Graduation, 1976-84

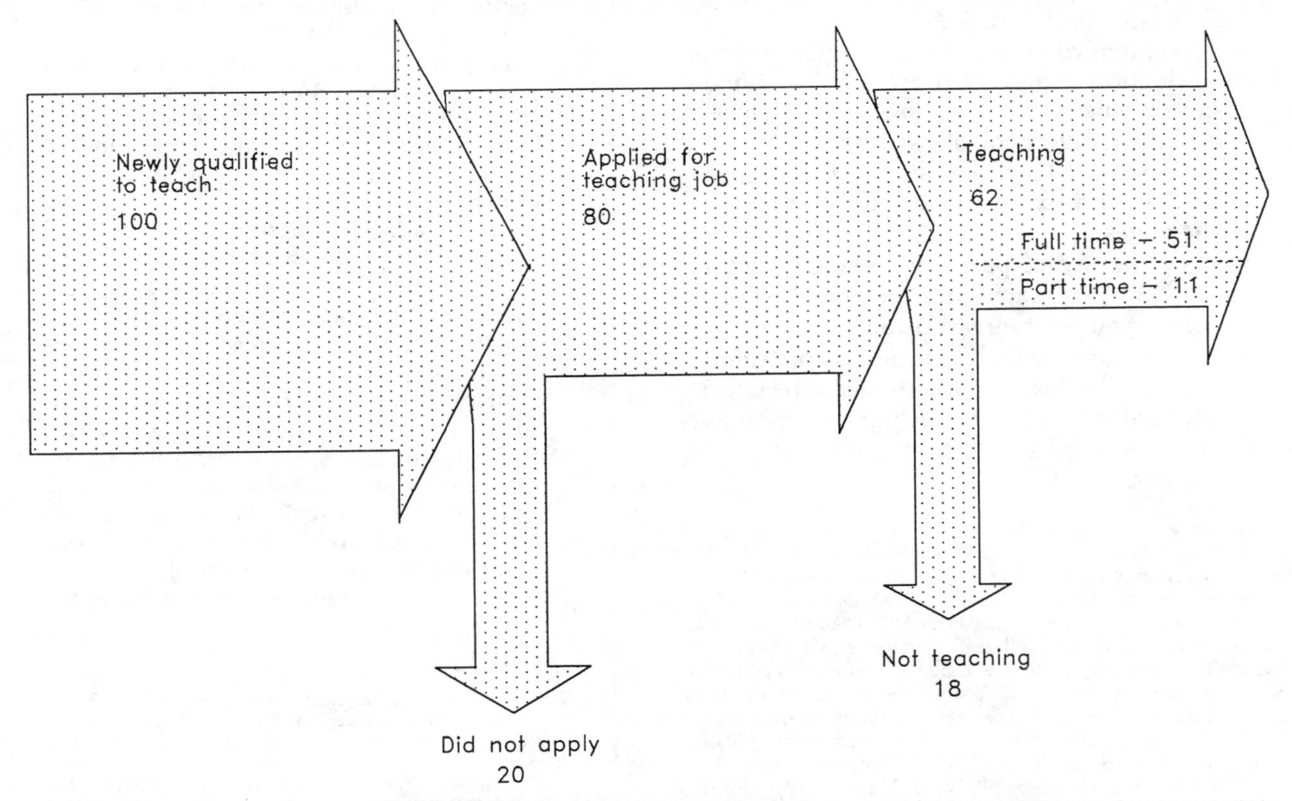

NOTE: "Newly qualified teachers" are defined as those graduates who were eligible for a teaching certificate or who were teaching at the time of the survey (and who had never taught full time before receiving this bachelor's degree). Bachelor's graduates are surveyed about 1 year after graduation.

SOURCE: Office of Technology Assessment, 1988; based on combined data from U.S. Department of Education, "Recent College Graduate Surveys of 1976-77, 1979-80, and 1983-84," unpublished data. Career pattern is similar in all years.

there will be a shortage, what its effects might be, or what the special aspects of the problem will be for mathematics and science teaching. Some aspects include:

- How intensified competition for new science and engineering baccalaureates will reduce the already small incentives for new graduates to consider mathematics and science teaching careers.
- The extent to which mathematics and science teachers who are already qualified but working in other occupations could be lured back to classroom teaching in response to higher salaries or changes in working conditions.
- The extent to which working and retired scientists and engineers could be retrained to enter the teaching work force. Several programs are attempting to retrain such people.
- How attrition of existing teachers (currently between 5 and 10 percent annually) can be reduced.[8] (See box 3-A.)
- The extent to which changes, such as the introduction of less restrictive certification requirements, the use of uncertified teachers, and the assignment of existing teachers out of their main teaching fields to teach mathematics and science classes, could cover shortages.
- How the use of part-time teachers, master teachers, or teaching assistantships could compensate for any shortages.
- The extent to which greater use of technologies, including computers, video recorders, and distance learning techniques, could reduce the need for mathematics and science teachers.[9]

[8]A survey of teacher attrition, based on followups of the National Longitudinal Survey of 1972, is in Barbara Heyns, "Educational Defectors: A First Look at Teacher Attrition in the NLS-72," *Educational Researcher*, April 1988, pp. 24-32. One surprising finding of this and other studies (such as the U.S. Department of Education's Survey of Recent College Graduates) is that a large number of those who complete teacher training programs never, in fact, teach. In the 9 years between 1977 and 1986, one-quarter of those qualified never taught, and 40 percent of those who became newly qualified teachers in 1983-84 had not become teachers by 1985. See also Richard J. Murnane, "Understanding Teacher Attrition," *Harvard Educational Review*, vol. 57, No. 2, May 1987, pp. 177-182, which finds that chemistry and physics teachers in Michigan in the 1970s were likely to leave teaching faster than were biology and history teachers.

[9]There is no evidence that technology replaces teachers. The use of satellite, cable, and other telecommunications technologies en-

Box 3-A.—Reasons Why Physics Teachers Leave High School Teaching

A 1983 survey reported some of the reasons why physics teachers leave teaching.[1] Those with a graduate degree in physics can readily find well-paying jobs in industry; either they never enter the teaching profession or they hastily depart. In general, the survey found, physics teachers leave for the following reasons:

- Instructional laboratories are poorly equipped and budgets are inadequate for making improvements.
- It is difficult to remain professionally active. There are seldom funds for teachers to attend professional meetings, to keep up-to-date with scientific literature and advances, or to meet and share experiences with teachers in other schools. This feeds a sense of isolation.
- Accountability to local, State, and Federal bodies has multiplied both teacher paperwork and administrative duties.
- There is a lack of identification by most school administrators with the problems that interfere with quality science teaching. School administrators, the survey reports, are often not interested in improving science teaching.
- There is a lack of respect within the local community. Like teachers of all subjects, physics teachers are often criticized in school board meetings as being greedy and inefficient, particularly when funding decisions are made.
- Voters do not support the schools, as evidenced by the willingness to vote down school bond issues in the early-1980s, even at the expense of reductions in the size and quality of the teaching work force. This strong pressure to cut taxes is especially evident in smaller communities whose demographics favor needs other than those of the student population.

[1]For reasons why physics teachers leave teaching, see Beverly Fearn Porter and William H. Kelly, "Why Physicists Leave Teaching," *Physics Today*, September 1983, pp. 32-37. Also see American Association of Physics Teachers, *The Role, Education, and Qualifications of the High School Physics Teacher* (College Park, MD: AAPT Committee on Special Projects for High School Physics, 1988). As this technical memorandum went to press, the American Institute of Physics released a new report, Michael Neuschatz and Maude Covalt, *Physics in the High Schools: Findings From the 1986-87 Nationwide Survey of Secondary School Teachers of Physics* (New York, NY: American Institute of Physics, 1988).

Some secondary school principals are having difficulty hiring science teachers and (to a lesser extent) mathematics teachers. The 1985-86 National Survey of Science and Mathematics Education found that 70 percent of secondary school principals said that they were having difficulty hiring physics teachers, 60 percent were having difficulty with chemistry and computer science teachers, and over 30 percent were having difficulty locating biology and life sciences teachers. The survey found that few schools had incentive programs to attract teachers to shortage fields; retraining programs are the more common method of supplying shortage fields.[10]

After years of declining interest among college freshmen in becoming teachers, there has been an upturn since 1986.[11] A 1985-86 survey estimated that about 20 percent of science and mathematics teachers are expected to retire in the next decade.[12] The result of these opposite trends is anybody's guess, so speculations abound.

Salaries

Many policymakers and educators point to the generally low level of teachers' salaries and claim that neither the number nor the quality of mathematics and science teachers can be improved until these salaries are increased substantially.[13] In fact, teachers' salaries are rising. In real terms, average annual public school teacher salaries fell during the 1970s by about 10 percent from their all-time high in the early 1970s. By 1984-85, they had risen to just under what they were in 1969-70. The mean teacher salary in 1986 was about $25,000, but with large variations among the States.[14] The effects of these increases on teacher supply and quality, which take time to show up, may yet be very positive. Already, there is some increased interest among college freshmen in teaching careers.

The attractiveness of different occupations to new college graduates is shaped by the immediate starting salaries as well as prospective long-term earnings. Students with considerable debts from their baccalaureate education, it is argued, need a substantial source of income to start paying off these debts. Starting teaching salaries have consistently been lower than those in other professions, and have not increased as rapidly during the last decade. (See figure 3-2.)

A particular controversy for mathematics and science teachers is whether they should be paid more than other teachers in order to attract people to fill shortages. A recent survey indicated that a majority of secondary mathematics and science teachers would support differential pay of this kind, and many principals are also in favor of this. Support among those who teach mathematics and science at the elementary level is weaker. Traditionally, teacher unions have argued that teachers should be paid the same, regardless of their subject specialization.[15]

Minority Teachers

Because of the declining proportion of Blacks and Hispanics entering college and because of the expanded career options now open to them, the

ables school districts to provide instruction from one site to many sites—but teachers are not replaced. Instead, these distance learning projects aggregate sparsely populated classrooms of two or three students to more "regular" sized classrooms (Linda Roberts, Office of Technology Assessment, personal communication, September 1988).

[10]Iris R. Weiss, *Report of the 1985-86 National Survey of Science and Mathematics Education* (Research Triangle Park, NC: Research Triangle Institute, November 1987), tables 72, 73.

[11]For the recent upturn in college freshmen interest in education majors, see Robert Rothman, "Proportion of College Freshmen Interested in a Career in Teaching Up, Survey Finds," *Education Week*, vol. 7, Jan. 20, 1988, pp. 1, 5. Eight percent of 1987 college freshmen planned teaching careers, up from 4.7 percent in 1982, but well below the 20 percent level in the early 1970s. The number of physics baccalaureates entering teaching also increased from only 23 in 1981 to about 100 in 1986 (of a total of 5,214 physics degree recipients in 1986). *Physics Today*, "Survey of Physics Bachelors Finds That More Plan to Teach," September 1987, p. 76.

[12]Weiss, op. cit., footnote 10, p. 64, table 36.

[13]Salaries are important, but are not the only factor that affects whether teachers enter or remain in teaching. Working conditions and the wider societal perception of the value of school teaching are also important influences. See, for example, Russell W. Rumberger, "The Impact of Salary Differentials on Teacher Shortages and Turnover: The Case of Mathematics and Science Teachers," *Economics of Education Review*, vol. 6, No. 4, 1987, pp. 389-399. Rumberger finds that the disparity between engineering and mathematics/science teaching salaries has some effect on teacher shortages and turnover; the disparity, however, offers far less than a complete explanation.

[14]For example, between 1969-70 and 1984-85, Alaska teacher salaries dropped by 34 percent in real terms, whereas those in Wyoming and Texas rose by 14 percent. U.S. Department of Education, Office of Educational Research and Improvement, Center for Education Statistics, *Digest of Education Statistics 1987* (Washington, DC: U.S. Government Printing Office, May 1987), tables 51-53.

[15]Weiss, op. cit., footnote 10, table 74.

Figure 3-2.—Starting Salaries for Teachers, Compared to Other Baccalaureates in Industry, 1975-87 (constant 1987 dollars)

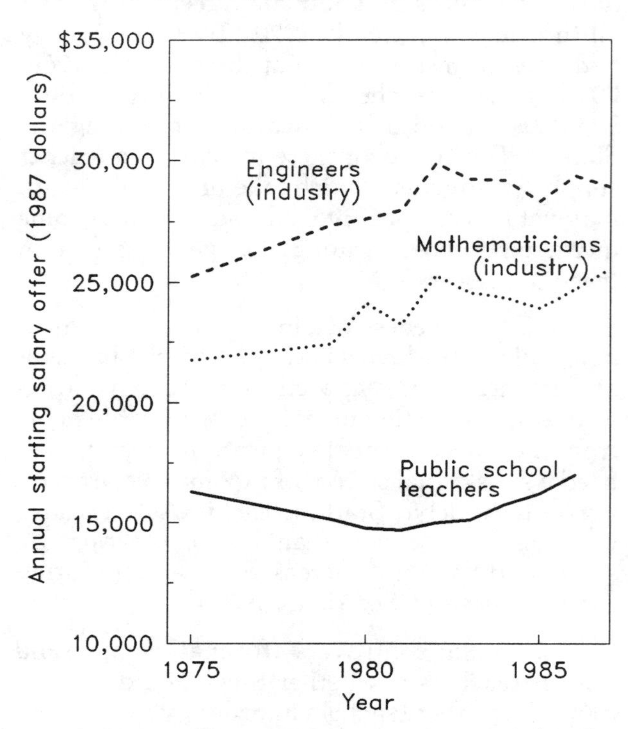

NOTE: Uses gross national product deflator. Industry estimates of salary offers are based on a survey of selected companies; this may tend to inflate salaries slightly. The teacher data are "minimum mean" salary, from the National Education Association, and probably are underestimates. Various other salary surveys report slightly different data. However, the basic message remains the same: teachers are paid much less than most other baccalaureates.

SOURCE: U.S. Department of Commerce, *Statistical Abstract of the United States* (Washington, DC: U.S. Government Printing Office, 1988), p. 130; based on data from The Northwestern Endicott-Lindquist Report, Northwestern University.

number of minorities electing teaching careers is declining. There are, in the first instance, comparatively few Black or Hispanic mathematics and science teachers. Data from a recent survey (see table 3-1) indicate that the majority of Black teachers are in the elementary grades; only 3 percent and 5 percent, respectively, of mathematics and science teachers in grades 10 to 12 are Black. Only 1 percent of teachers of both these subjects are Hispanic. For now, the proportion of minorities in the teaching force is increasing slightly, but several commentators warn of future shortages of minority teachers, particularly in mathematics and science. Such a shortage poses particular problems to schools with high minority enrollments, denying minority children role models (among other things that minority teachers provide). Making higher education more attractive and attainable for future Black and Hispanic teachers will help increase the supply of the minority teaching force.[16]

Table 3-1.—Mathematics and Science Teachers, by Race and Ethnicity, 1985-86 (in percent)

Subject and grade	Black	Hispanic	White	Other[a]	Unknown
Mathematics					
K-3	10	1	84	0	4
4-6.......	12	2	84	0	2
7-9.......	6	1	90	1	3
10-12.....	3	1	94	1	1
Science					
K-3	9	4	82	1	4
4-6.......	8	4	86	1	1
7-9.......	5	1	88	1	4
10-12.....	5	1	92	2	1

[a]Includes Native American, Alaskan Native, Asian, and Pacific Islander.
NOTE: Some rows do not sum to 100 percent due to rounding.

SOURCE: Iris R. Weiss, *Report of the 1985-86 National Survey of Science and Mathematics Education* (Research Triangle Park, NC: Research Triangle Institute, November 1987), table 35.

Certification and Misassignment of Mathematics and Science Teachers

Each State sets specifications, designed to ensure a minimum level of professional competence, for the academic preparation of teachers. These specifications, which take the form of requirements for a minimum number and combination of college-level courses in mathematics, science, and education, are enforced through certification and periodic recertification of teachers. Certification requirements vary considerably from State to State (see table 3-2), and there are differences in the extent to which they are enforced. The States may also require examinations, such as the National Teachers' Examination, for either initial certification or later recertification.[17]

[16]Shirley M. McBay, *Increasing the Number and Quality of Minority Science and Mathematics Teachers* (New York, NY: Carnegie Forum on Education and the Economy, Task Force on Teaching as a Profession, January 1986); Patricia Albjerg Graham, "Black Teachers: A Drastically Scarce Resource," *Phi Delta Kappan*, April 1987, pp. 598-605; and Blake Rodman, "AACTE Outlines Plan to Recruit Minorities Into Teaching," *Education Week*, Jan. 13, 1988, p. 6.

[17]At the elementary level, most teachers are certified as elementary teachers without particular specialization, but, at the secondary level, some specialization in certification is the norm. About one-half of the States license secondary teachers to teach in any science subject, while others restrict certification to a particular field,

(continued on p. 60)

Table 3-2.—Mathematics and Science Teacher Certification Requirements by State, June 1987

	Course credits by certification field						
	Math	Science, Broad-field	Biology, Chemistry, Physics	Earth Science	General science	Teaching methods: science/math	Supervise teaching experience
Alabama	27	52	27	27	27	Yes	9
Alaska	None	None	None	None	None	No	None
Arizona	30	30	30	30	30	Yes	8
Arkansas	21		24	24	24	No	12 wks
California	45		45			No	[a]
Colorado	[b]	[b]	[b]	[b]	[b]	Yes	400 hrs
Connecticut	18		18	18	21	No	6
Delaware	30		39-45	39	36	Yes	6
District of Columbia	27	30	30	30	30	Yes	1 sem
Florida	21		20	20	20	Yes(S)	6
Georgia (effective 9/88)	60 qtr	45 qtr	40 qtr	40 qtr		Yes(M)	15 qtr
Guam	18	18				No	None
Hawaii	[b]		[b]	[b]	[b]	[b]	[b]
Idaho	20	45	20	20		No	6
Illinois	24	32	24	24		Yes	5
Indiana	36		36	36	36	Yes	9 wks
Iowa	24	24	24	24	24	Yes	Yes
Kansas	[b]	[b]	[b]	[b]	[b]	[b]	[b]
Kentucky	30	48	30	30		No	9-12
Louisiana	20		20	20	32	No	9
Maine	18	18				Yes	6
Maryland	24	36	24	24	36	Yes	6
Massachusetts	36	36	36	36	36	Yes	300 hrs
Michigan	30	36	30	30		No	6
Minnesota	[c]	[c]	[c]	[c]	[c]	[c]	1 qtr
Mississippi	24		32	32	32	Yes(S)	6
Missouri	30	30	20	20	20	Yes	8
Montana	30 qtr	60 qtr	30 qtr	30 qtr	30 qtr	Yes	10 wks
Nebraska	30	45	24	24		Yes	320 hrs
Nevada	16	36	16	16	16	No	8
New Hampshire	[b]	[b]	[b]	[b]	[b]	[b]	[b]
New Jersey	30	30	30	30	30	No	[c]
New Mexico	24	24	24	24	24	Yes	6
New York	24		36	36	36	No	Yes
North Carolina	[c]	[c]	[c]	[c]	[c]	Yes	6
North Dakota	[c]	[c]	[c]	[c]	[c]	Yes	8
Ohio	30	60	30	30	30	Yes	[a]
Oklahoma	40	40	40	40	40	No	12 wks
Oregon	21	45	45	45	45	Yes(M)	15 qtr
Pennsylvania	[b]	[b]	[b]	[b]	[b]	[b]	[b]
Puerto Rico	30	30	30		30	Yes	3(S)5(M)
Rhode Island	30	30	30		30	Yes	6
South Carolina	24	30	12		18	Yes(M)	6
South Dakota	18	21	12	12	18	No	6
Tennessee	36 qtr	48 qtr	24 qtr	24 qtr	24 qtr	Yes	4
Texas	24	48	24	24		No	6
Utah	[c]	[c]	[c]	[c]	[c]	Yes	12
Vermont	18	18	18	18	18	Yes	None
Virginia	27		24	24	30	No	6
Virgin Islands	24	NA	NA	NA	NA	No	Yes
Washington	24	51	24	24		No	15
West Virginia	[c]	[c]	[c]	[c]	[c]	[c]	[c]
Wisconsin	34	54	34	34	34	Yes	5
Wyoming	24	30	12	12	12	No	1 course

KEY: Course credits = semester credit hours, unless otherwise specified; qtr = quarter credit hours; M = mathematics only; S = science only; NA = not available; blank space = no certification offered.

[a]1 semester full time or 2 semesters half time—California; supervised teaching experience and 300 hours clinical/field-based experience—Ohio.

[b]Certification requirements determined by degree-granting institution or approved competency-based program.

[c]Major or minor—North Dakota, Utah; 20 to 40 percent of program—Minnesota and North Carolina; courses matched with job requirements—West Virginia.

SOURCE: Rolf Blank and Pamela Espenshade, *State Education Policies Related to Science and Mathematics* (Washington, DC: Council of Chief State School Officers, State Education Assessment Center Science and Mathematics Indicators Project, November 1987), table 4.

Photo credit: William Mills, Montgomery County Public Schools

There are few minority teachers in mathematics and science to serve as role models for Black, Hispanic, and Native American children.

As part of the education reform movement, policymakers have tightened certification standards in the hope of raising the quality of teaching. Altering certification requirements may be an easy control on the system for policymakers to enact, but have little effect on actual classroom practices and teaching quality. However, some teachers teach without certification, either because they are new to the State and are working to achieve accreditation (and are teaching on an "emergency" basis) or because they are teaching subjects other than those which they are certified to teach.[18] An increasing number of science

(continued from p. 58)
such as physics. In each case, typical requirements are 24 to 36 semester-hours of college-level science courses. Ken Mechling, "Science Teacher Certification Standards: An Agenda for Improvement," *Redesigning Science and Technology Education: 1984 Yearbook of the National Science Teachers Association*, Rodger W. Bybee et al. (eds.) (Washington, DC: National Science Teachers Association, 1984), pp. 157-161.

[18]Data on the extent to which "uncertified" teachers are in charge of mathematics and science classes are fragmented and often inconsistent. Analysts differ on the interpretation of uncertified: sometimes the term is interpreted as including those without any kind of certification, sometimes it includes teachers who are certified but are teaching out of their main field of competence or certification (the two are not always the same), and other times it is used to include teachers who have provisional or emergency certification, but not full certification. (To the extent that there is great flexibility to

teachers, in particular, appear to be teaching subjects that they are either not licensed or not qualified to teach. A 1986 survey of 39 States (enrolling 28 percent of the student population) estimated that between 6 and 15 percent of all science teachers were uncertified in the field they were hired to teach. Biology had the lowest proportion of uncertified teachers, while earth and general science had the highest. About 8 percent of mathematics teachers were uncertified in that field.[19] The proportion of uncertified mathematics and science teachers was greatest in the Southeast region of the country. A 1985-86 survey indicates that as many as 20 percent of science teachers in grades 10 to 12 are not certified to teach the courses they are teaching: 4 percent are not certified at all, 6 percent have provisional certification, and 5 percent are certified in other fields (the remainder are presumably those certified in one science subject but teaching another). This same survey found that, of teachers of mathematics in grades 10 to 12, 4 percent were not certified at all and 4 percent had only provisional certification, while 10 percent were certified in fields other than mathematics. In total, 14 percent of these teachers were teaching courses that they were uncertified to teach.[20]

National data from the National Science Teachers Association (NSTA) indicate that the notion that a high school science teacher teaches only one science is increasingly a myth. And many science teachers teach mathematics or nonscience subjects as well. On average, about 8 percent of the course assignment of secondary science teachers is in mathematics, and 5 percent is in nonscience subjects. For example, about half of the teaching load of chemistry teachers is in chemistry, 12 percent in biology, and 15 percent in physics and general physical science.[21]

This pattern is reflected in the teaching of all subjects at the secondary level. The National Education Association estimates that 83 percent of all subject specialist secondary teachers devote all their teaching time to teaching the field that was their college major; 7 percent spend between 50 and 100 percent of their time in that field; and only 10 percent spend less than 50 percent of their time teaching in that field.[22]

While States condemn teaching without adequate certification, critics of the system of certification note that States tacitly condone it by permitting waivers of requirements and by failing to enforce certification requirements.[23] To the extent that shortages exist, States, school districts, or principals must choose whether it would be better to have a poorly qualified teacher teaching a science class than to have no teacher and no class at all.

A number of States have developed alternative certification routes for mature entrants to the teaching profession, particularly those who are already qualified scientists, engineers, or technicians. These programs focus particularly on recruiting mathematics and science teachers. A recent study estimates that there are 26 such programs nationally, and some have attracted Federal funding.[24]

issue such certification, States and school districts have an easy way to rectify any concerns about the number of uncertified teachers in the classroom.) Principals reportedly prefer often to retain existing uncertified teachers in classes where they have developed rapport with the class than introduce new, inexperienced, but fully certified teachers who would have much more difficulty teaching the class.

[19]Joanne Capper, *A Study of Certified Teacher Availability in the States* (Washington, DC: Council of Chief State School Officers, February 1987). These data are drawn from State needs assessments, mandated under Title II of the Education for Economic Security Act of 1984; the data analysis was funded by the National Science Foundation.

[20]Weiss, op. cit., footnote 10, table 46.

[21]Bill G. Aldridge, "What's Being Taught and Who's Teaching It," *The Science Curriculum: The Report of the 1986 National Forum for School Science*, Audrey B. Champagne and Leslie E. Hornig (eds.) (Washington, DC: American Association for the Advancement of Science, 1987), pp. 207-223.

[22]National Education Association, *Status of the American Public School Teacher 1985/86* (West Haven, CT: National Education Association, 1987), table 18. These data are based on a definition of misassignment as teachers assigned outside their main college preparatory field. This is an imperfect measure, because some teachers are qualified to teach in subjects that were not their college major.

[23]American Federation of Teachers/Council for Basic Education, *Making Do in the Classroom: A Report on the Misassignment of Teachers* (Washington, DC: 1985); Aldridge, op. cit., footnote 21, 1985, p. 84.

[24]These programs enjoy some success, but data on their impact are very limited. Anecdotal evidence suggests that those who make such transitions are not likely to be the best and the brightest in their fields of origin, but there is no way (yet) of judging their quality relative to teachers in the field they have joined. See Linda Darling-Hammond and Lisa Hudson, Rand Corp., "Precollege Science and Mathematics Teachers: Supply, Demand, and Quality,"

(continued on next page)

These programs look promising, and could be expanded in the interests both of the quantity and quality of the entry-level science and mathematics work force. New York City has a program to relicense teachers of subjects other than mathematics and science in these fields. (See box 3-B.)

(continued from previous page)
mimeo, 1987, p. 51; Shirley R. Fox, *Scientists in the Classroom: Two Strategies* (Washington, DC: National Institute for Work and Learning, 1986); and Nancy E. Adelman et al., *An Exploratory Study of Teacher Alternative Certification and Retraining Programs* (Washington, DC: Policy Studies Associates, Inc., October 1986).

In an interesting initiative in Hammond, Indiana, a chemistry teacher works part time in a local steel company and part time in the local high school. His salary is shared by the school district and the company, and some of his classes are taught in the industrial research laboratory. This arrangement originated in the enthusiasm of the teacher and the local community, and could be replicated elsewhere.[25]

[25]Brent Williamson, high school teacher, personal communication, February 1988.

Box 3-B.—Mathematics and Science Relicensing Board

During the summer of 1984, the New York City Board of Education decided to retrain teachers of other subjects to teach mathematics and science.[1] The board realized that urban schools were most vulnerable to any future teacher shortage, because the staffs of such schools are typically older than their suburban counterparts and because teachers prefer to work in suburban school systems. On average, the proportion of New York's teaching force that will retire soon is about twice that of the national teacher work force. The Mathematics and Science Relicensing Board's task is to "re-tool" teachers of subjects such as English and history in which there is a teacher surplus, and, by April 1987, had succeeded in awarding 400 licenses in mathematics and science. These recertified teachers are primarily female and minority; data from 1985 indicate that 61 percent of program participants were women and 45 percent were Black or Hispanic, percentages that were considerably higher than those of the teaching force of New York City as a whole (approximately 22 percent of secondary school teachers of all subjects are members of minority groups and 50 percent are female). The participants think that their new licenses will make them more mobile and give them the challenge of working in a new field.

The recertification program was free to the teachers, and took place after hours and in the summer at nearby universities and colleges. These classes were arranged especially for the Relicensing Board, and cost between $45 to $70 per credit, depending on the class and university. The academic program is similar to an undergraduate major, and teachers can be simultaneously certified in different subjects, as well as at different levels. After coursework is completed, candidates must pass a city-wide examination, and, after passing, take a probationary position as a teacher in the new field. After 2 years, tenure is reviewed. If granted, the licensees are certified both for course and probationary period completion. Seminars and tutorials are also offered to assist in preparation.

The program benefits from its cooperative model; it is the only one in the United States that brings together so many different institutions and groups with an interest in staffing city classrooms. Funds come from the city, advertising is by courtesy of the United Federation of Teachers, the Board of Education approves the program, the Division of Curriculum and Instruction establishes and manages the program, and 11 universities and colleges take part. Everyone benefits, particularly the academic institutions that have a strong interest in improving the level of preparation of their future students.

An important benefit of the board has been that the large number of Black, Hispanic, and female teachers trained through the program has increased the interest of their minority and female students in mathematics and science. An evaluation of the program is in progress.

[1]Bruce S. Cooper, "Retooling Teachers: The New York Experience, *Phi Delta Kappan*, vol. 68, No. 8, April 1987, pp. 606-609.

THE PROFESSIONAL STATUS OF THE TEACHING WORK FORCE

Concern about teacher shortages and quality comes at a time when the teaching profession as a whole feels embattled and undervalued, but also recognizes its key role in education and in education reform. Seemingly endless commission reports have cited the need to give greater status, more recognition, and higher salaries to teachers.[26] Although teachers aspire to belong to a profession, few feel that they truly do. Many argue that administrators and school boards, not teachers, define standards of conduct in schools, teaching methods, and curricula. Teachers are constrained by many rules and regulations, many of which conflict with each other and which, taken together, sap the enthusiasm of many teachers. In some ways, the process of increasing requirements and paperwork is a kind of "de-skilling" of the teaching work force: the skill of teaching is removed from teachers and given to those who make and enforce the "rules."[27]

One way of redressing the balance is to give teachers more say in setting the professional qualifications and standards for membership in the teaching force. The recommendation of the Carnegie Task Force on Teaching as a Profession for a national certification board is being implemented; its first members were nominated in May 1987. Eventually, such certification might replace State certification. Parallel moves are afoot in the mathematics and science teaching profession. In 1984, NSTA estimated that about 30 percent of all secondary mathematics and science teachers were either "completely unqualified or severely underqualified" to teach these subjects.[28] NSTA launched its own certification program for science teachers in October 1986. The fact that many "single-subject" teachers teach a good deal of other sciences and mathematics has led NSTA to devise a two-track secondary certification program: one for general science teaching, the other for single-subject science teaching. Currently, fewer than one-third of the science teaching force would meet NSTA's standards.[29] The guidelines of both NSTA and the National Council of Teachers of Mathematics (NCTM, set in 1981) are listed in table 3-3.

The Quality of Mathematics and Science Teachers

Mathematics and science teacher quality is not easily measured.[30] There are three related and commonly used indicators of teacher quality: possession of State certification, conformity to guidelines established by such bodies as NSTA and NCTM, and the amount of college-level coursework that teachers have had (on which the other two indicators are based). Many commentators caution against equating course preparation with teacher quality. Nevertheless, reliable data exist only for this measure and it is the one used here, along with teachers' own perceptions of their confidence and abilities.[31]

---

[26] See, for example, Task Force on Teaching as a Profession, *A Nation Prepared: Teachers for the 21st Century* (New York, NY: Carnegie Forum on Education and the Economy, 1986); National Commission for Excellence in Teacher Education, *Call for Change in Teacher Education* (Washington, DC: American Association of Colleges for Teacher Education, 1985); National Science Board, Commission on Precollege Education in Mathematics, Science, and Technology, *Educating Americans for the 21st Century* (Washington, DC: 1983); and Paul E. Peterson, "Did the Education Commissions Say Anything?" *The Brookings Review*, winter 1983, pp. 3-11. See also The Carnegie Foundation for the Advancement of Teaching, *Report Card on School Reform: The Teachers Speak* (Washington, DC: 1988), which characterizes recent reforms as involving greater regulation of easily manipulated elements of education (such as graduation requirements, testing for minimum competency, requirements on teacher preparation, and tester testing) rather than renewal. Teachers have largely not been involved in these reforms, only ordered to undertake them. Nearly one-half of teachers report that morale in teaching has actually fallen since 1983, when the current wave of reforms began.

[27] Martin Carnoy and Henry M. Levin, *Schooling and Work in the Democratic State* (Stanford, CA: Stanford University Press, 1985), pp. 157-158.

[28] K.L. Johnston and B.G. Aldridge, "The Crisis in Science Education: What Is It? How Can We Respond?" *Journal of College Science Teaching* (September/October 1984), quoted in National Science Board, op. cit., footnote 6, p. 37.

[29] John Walsh, "Teacher Certification Program Under Way," *Science*, vol. 235, Feb. 20, 1987, pp. 838-839; and Robert Rothman, "Science Teachers Laud Certification Program, But Few Seen Qualified," *Education Week*, Apr. 8, 1987, pp. 6, 10.

[30] See, generally, Darling-Hammond and Hudson, op. cit., footnote 24.

[31] Rolf K. Blank and Senta A. Raizen, National Research Council, "Background Paper for a Planning Conference on a Study of Teacher Quality in Science and Mathematics Education," unpublished working paper, April 1985. Unfortunately, few people seem to have asked the consumers of this teaching, the students, what they think of their teachers' abilities. Better ways of measuring quality might be either to observe teachers' performance in the class-

(continued on next page)

Table 3-3.—Guidelines for Mathematics and Science Teacher Qualifications Specified by the National Council of Teachers of Mathematics (NCTM) and the National Science Teachers Association (NSTA)

NCTM guidelines

Early elementary school

The following three courses, each of which presumes a prerequisite of 2 years of high school algebra and 1 year of geometry:

1. number systems
2. informal geometry
3. mathematics teaching methods

Upper elementary and middle school

The following four courses, each of which presumes a prerequisite of 2 years of high school algebra and 1 year of geometry:

1. number systems
2. informal geometry
3. topics in mathematics (including real number systems, probability and statistics, coordinate geometry, and number theory)
4. mathematics methods

Junior high school

The following seven courses, each with a prerequisite of 3 to 4 years of high school mathematics, beginning with algebra and including trigonometry:

1. calculus
2. geometry
3. computer science
4. abstract algebra
5. mathematics applications
6. probability and statistics
7. mathematics methods

Senior high school

The following 13 courses, which constitute an undergraduate major in mathematics, and which each presume a prerequisite of 3 to 4 years of high school mathematics, beginning with algebra and including trigonometry:

1-3. three semesters of calculus
4. computer science
5-6. linear and abstract algebra
7. geometry
8. probability and statistics
9-12. one course each in: mathematics methods, mathematics applications, selected topics, and the history of mathematics
13. at least one additional mathematics elective course

NSTA standards

Elementary level

1. Minimum 12 semester-hours in laboratory- or field-oriented science including courses in biological, physical, and earth sciences. These courses should provide science content that is applicable to elementary classrooms.
2. Minimum of one course in elementary science methods (approximately 3 semester-hours) to be taken after completion of content courses.
3. Field experience in teaching science to elementary students.

Middle/junior high school level

1. Minimum 36 semester-hours of science instruction with at least 9 hours in each of biological or earth science, physical science, and earth/space science. Remaining 9 hours should be science electives.
2. Minimum of 9 semester-hours in support areas of mathematics and computer science.
3. A science methods course designed for the middle school level.
4. Observation and field experience with early adolescent science classes.

Secondary level

General standards for all science specialization areas:

1. Minimum 50 semester-hours of coursework in one or more sciences, plus study in related fields of mathematics, statistics, and computer applications.
2. Three to five semester-hour course in science methods and curriculum.
3. Field experiences in secondary science classrooms at more than one grade level or more than one science area.

Specialized standards

1. Biology: minimum 32 semester-hours of biology plus 16 semester-hours in other sciences.
2. Chemistry: minimum 32 semester-hours of chemistry plus 16 semester-hours in other sciences.
3. Earth/space science: minimum 32 semester-hours of earth/space science, specializing in one area (astronomy, geology, meteorology, or oceanography), plus 16 semester-hours in other sciences.
4. General science: 8 semester-hours each in biology, chemistry, physics, earth/space science, and applications of science in society. Twelve hours in any one area, plus mathematics to at least the precalculus level.
5. Physical science: 24 semester-hours in chemistry, physics, and applications to society, plus 24 semester-hours in earth/space science; also an introductory biology course.
6. Physics: 32 semester-hours in physics, plus 16 in other sciences.

SOURCE: National Council of Teachers of Mathematics and the National Science Teachers Association.

The national teaching force has good credentials; over 50 percent of *all* teachers now have at least a master's degree.[32] A 1985-86 survey found, in grades 10 to 12, that 63 percent of science teachers and 55 percent of mathematics

(continued from previous page)

room or to evaluate the outcomes of teaching through the progress made by students (which is becoming more common as States upgrade course requirements for high school graduation).

[32]National Education Association, *Status of the American Public School Teacher 1985-86* (West Haven, CT: 1987), tables 1-2.

teachers had earned degrees beyond the baccalaureate. The same survey also found that 40 percent of mathematics teachers had degrees in mathematics, and 60 percent of science teachers had degrees in a science field.[33] By contrast, only 1 to 2 percent of mathematics and science teachers at the elementary school level had degrees in these fields.

The National Survey of Science and Mathematics Education gathered its data from about 4,500 teachers from all grades in 1985-86.[34] The survey showed that many elementary mathematics and science teachers have taken very few college-level courses in these subjects, while secondary teachers of these subjects have much more extensive preparation. (See tables 3-4 and 3-5.)

The survey indicated that over 85 percent of elementary science teachers have taken at least one course in methods for teaching elementary school science, and about 90 percent have taken at least one college-level science course (typically biology, psychology, or physical science).[35] However, although 90 percent of elementary mathematics teachers have taken at least one course in methods for teaching mathematics, only about 40 percent have taken at least one college-level mathematics class. Most have taken instead mathematics courses especially designed for elementary mathematics teachers. Elementary school teachers feel good about mathematics; 99 percent feel well-qualified to teach it, compared to 64 percent who feel well-qualified to teach science, particularly physical science. About 80 percent of elementary mathematics and science teachers enjoy teaching these subjects. Inservice training is also not reaching many elementary teachers; more than 40 percent report that they have had no inservice training in the last year, and another 25 percent have had less than 6 hours in total during the year.

About 90 percent of junior high and high school mathematics and science teachers have taken at least introductory biology in college, over 70 percent have taken general physics, 50 percent geology, and 80 percent general chemistry. Many have specialized in biology and life sciences; few have specialized in physical sciences. About one-half have taken at least eight courses in life science, but only 14 percent of them have had eight courses in chemistry, and 10 percent eight courses in physics and earth sciences. As a group, over 90 percent of secondary science teachers enjoy teaching science, although 35 percent think that science is a difficult subject to learn.[36]

Table 3-4.—College-Level Courses Taken by Elementary and Secondary Mathematics Teachers

	Percentage of teachers with course[b]			
	Elementary		Secondary	
Course titles[a]	K-3	4-6	7-9	10-12
General Methods of Teaching	94	93	90	93
Methods of Teaching Elementary School Mathematics	90	90		
Methods of Teaching Middle School Mathematics	14	27	37	25
Methods of Teaching Secondary School Mathematics			53	80
Supervised Student Teaching	82	83	79	81
Psychology, Human Development	83	87	84	87
Mathematics for Elementary School Teachers	89	90		
Mathematics for Secondary School Teachers	11	21		
Geometry for Elementary or Middle School Teachers	17	21		
College Algebra, Trigonometry, Elementary Functions	30	37	80	87
Calculus	8	12	67	89
Advanced Calculus			39	63
Differential Equations			39	61
Geometry[c]	5	7	67	80
Probability and Statistics	21	27	59	76
Abstract Algebra/Number Theory			48	69
Linear Algebra			48	69
Applications of Mathematics/Problem Solving			34	39
History of Mathematics			26	37
Other upper division mathematics			37	63
Sample N =	433	246	671	565

[a]Omits courses in computer programming and instructional uses of computers.
[b]Empty cells mean data were not reported in original tabulation.
[c]Upper division geometry in case of elementary teachers.

SOURCE: Iris R. Weiss, *Report of the 1985-86 National Survey of Science and Mathematics Education* (Research Triangle Park, NC: Research Triangle Institute, November 1987), tables 40, 44.

[33]Weiss, op. cit., footnote 10, tables 38, 46.
[34]Ibid. For the higher grades, data are reported in two categories: teachers in grades 7 to 9 and grades 10 to 12; here they are summarized as averages for grades 7 to 12 combined.
[35]Ibid., tables 39-40.

[36]Ibid., tables 41, 44.

Table 3-5.—College-Level Courses Taken by Elementary and Secondary Science Teachers

Course titles[a]	Percentage of teachers with course[b]			
	Elementary		Secondary	
	K-3	4-6	7-9	10-12
General Methods of Teaching	95	95	94	94
Methods of Teaching Elementary School Science	87	88		
Methods of Teaching Middle School Science	7	20	30	20
Methods of Teaching Secondary School Science			61	82
Supervised Student Teaching	77	87	83	79
Psychology, Human Development	83	88	85	87
Biology, Environmental, Life Sciences	83	87		
Chemistry	30	37		
Physics	17	21		
Physical Science	58	61		
Earth/Space Science	39	51		
No science courses	5	5		
Only one science course	18	12		
Two science courses	40	40		
Life Sciences:				
Introductory Biology			91	85
Botany, Plant Physiology			70	73
Cell Biology			54	58
Ecology, Environmental Sciences			62	63
Genetics, Evolution			55	64
Microbiology			48	53
Physiology			63	65
Zoology, Animal Behavior			64	71
Chemistry:				
General Chemistry			76	92
Analytical Chemistry			30	47
Organic Chemistry			51	70
Physical Chemistry			21	32
Biochemistry			25	34
Physics:				
General Physics			73	81
Electricity and Magnetism			18	28
Heat and Thermodynamics			16	24
Mechanics			15	26
Modern or Nuclear Physics			12	23
Optics			11	18
Earth/Space Sciences:				
Astronomy			40	36
Geology			56	49
Meteorology			27	20
Oceanography			26	19
Physical Geography			39	25
Other:				
History of Science			21	23
Science and Society			18	16
Engineering			8	12
Sample N =	431	273	658	1,050

[a]Omits courses in computer programming and instructional uses of computers.
[b]Empty cells mean data were not reported in original tabulation.

SOURCE: Iris R. Weiss, *Report of the 1985-86 National Survey of Science and Mathematics Education* (Research Triangle Park, NC: Research Triangle Institute, November 1987), tables 39 and 41.

Of mathematics teachers in grades 7 to 12, over 80 percent have had at least college algebra, trigonometry, or elementary functions, and about 70 percent of them have had calculus. Still, about 7 percent feel inadequately qualified to teach mathematics, and over 25 percent had not taken a college course for credit in the last 12 years (55 percent during the last 5 years). Over 50 percent have not had more than 6 hours of inservice education during the last year. This translates into a lack of confidence in teaching skills. About 20 percent of elementary teachers felt very well-qualified to teach mathematics and science respectively; another 20 percent felt they were not well-qualified to teach science.[37]

Options for Improving the Quality of Mathematics and Science Teachers

More States indicate shortages of *quality* science and mathematics teachers than of teachers with appropriate qualifications to teach these subjects. Credentials are not enough. Most States have attempted to alleviate their shortages through higher teacher salaries, and some also use special loan and staff development programs for mathematics and science teachers in order to retain good teachers and retrain teachers from other fields. Iowa, for example, grants loans to current teachers to upgrade their skills in mathematics and science teaching, and sponsors summer training institutes. Idaho uses Title II funds to provide scholarships to potential science or mathematics teachers who want to be recertified in these subject areas.

At least 26 States have inservice teacher training programs for science and mathematics instructors, most involving loans or scholarships to promote additional coursework. The Teacher Summer Business Training and Employment Program in New York partly reimburses industry for science, mathematics, computer, or occupational education teachers employed by business and industry during the summer. In Kentucky, Title II funds support the Science Improvement Project in low-income districts with histories of low achievement.

[37]See National Science Board, op. cit., footnote 6, pp. 27-28.

About 10 States now offer alternative certification programs for prospective mathematics and science teachers. For example, Utah awards "Eminence Certificates" to qualified professionals such as engineers and doctors, which allows them to teach up to two classes per day. Other, more innovative means of recruiting new mathematics and science teachers include hiring teachers from overseas. (California and Georgia recruit science and mathematics teachers from the Federal Republic of Germany, and Kansas City, Missouri, has imported teachers from Belgium.) Florida holds an intensive teacher job fair each June, called "The Great Florida Teach-In," designed to attract and place new teachers.

The quality of the mathematics and science teacher work forces can be improved before people enter the classroom as teachers (generally referred to as preservice) or when they are actively teaching (inservice). Given the low labor turnover of the teaching force, between 5 and 10 percent each year in all subjects,[38] the way to upgrade teaching quality is via inservice programs. Yet there is considerable national anxiety about the perceived deficiencies of preservice teacher preparation in all disciplines.[39]

Preservice Education

While many talented people do become teachers, it is sometimes suggested that teacher education is not challenging.[40] Critics further charge that teacher preparation programs fail to make effective links between courses on mathematics and science and those on education, and therefore, teachers are unable to convert courses on classroom teaching techniques and theories of learning. In addition, such courses convey a simplistic view of science as a monolithic collection of facts, embodied in enormous textbooks, giving students a false impression of the nature of scientific inquiry.

Teachers agree that experiments and hands-on activities are more effective than book work, but feel the overriding need to cover material in encyclopedic fashion. The extensive use of factual recall tests creates incentives to cover the content, rather than process, of the subject matter. Thus, teacher preparation may be more telling than their classroom practice. In college, prospective teachers model their attitudes and teaching practices on those of their college professors and, indeed, on their own school teachers. They employ the teaching techniques, such as lectures and rote memorization, that they were either forced to suffer or benefited from when they were students. School district curriculum guides and testing fuel teachers' reliance on tools for covering concepts and facts, one by one, without drawing links and brightening the big picture of science. Preference may signal a lack of alternatives; teachers may have neither the tools nor the opportunity to become comfortable with them to change their approach.

Photo credit: William Mills, Montgomery County Public Schools

Most reports on reforming education single out the importance of improving the status, appeal, and quality of the teaching profession.

[38]National Education Association, op. cit., footnote 32, table 13; Blake Rodman, "Attrition Rate for Teachers Hits 25-Year Low, Study Finds," *Education Week*, Oct. 14, 1987, p. 8.

[39]For an overview, see Frank Ambrosie and Paul W. Haley, "The Changing School Climate and Teacher Professionalization," *NASSP [National Association of Secondary School Principals] Bulletin*, vol. 72, January 1988, pp. 82-89. The following two sections are based in part on Iris R. Weiss, OTA Workshop on Mathematics and Science Education K-12: Teachers and the Future, Summary Report, September 1987.

[40]National Science Board, op. cit., footnote 6, p. 25. As Bernard R. Gifford, Dean of the School of Education, University of California-Berkeley, puts it: "What's wrong with schools and departments of education today is very simple. Education suffers from congenital prestige deprivation." See Anne C. Roark, "The Ghetto of Academe: Few Takers (Teacher Colleges)," *Los Angeles Times*, Mar. 13, 1988, p. 6. A new book dissects the origins and repercussions of this prestige deprivation on university campuses. See Geraldine Joncich Clifford and James W. Guthrie, *Ed School. A Brief for Professional Education* (Chicago, IL: University of Chicago Press, 1988).

There is still no complete model of what the mathematics and science teacher curriculum should be. Simply requiring more mathematics and science courses for certification will not automatically improve teacher quality, given the content of these courses and the way they are often taught. The National Science Foundation (NSF), for example, has recently begun a program to develop new models for preservice preparation of middle school teachers.

One particular controversy in mathematics and science teacher education is whether future teachers should be expected to have a baccalaureate degree in a discipline plus some professional training. At present, many teachers at the elementary level earn baccalaureate degrees in education, but 97 percent of elementary mathematics teachers and 95 percent of elementary science teachers have a degree in subjects other than science or science education. At the high school level, however, 40 percent of mathematics teachers and 60 percent of science teachers have a degree in those subjects, and another 36 and 24 percent, respectively, have a degree in mathematics and science education or a joint degree in a mathematics and science subject and science and mathematics education.[41]

Several groups that have studied the future of the teaching profession in the current reform movement have looked at this issue. The Holmes Group (an informal consortium of education deans in research universities) has attached priority to upgrading elementary and secondary teachers' specific knowledge by insisting that they have a baccalaureate degree in a subject area. The Holmes Group has also called for much greater use of specialized teaching, and for more subject-intensive preparation of those teachers.[42] So far, only Texas has changed its certification requirements in this way; after 1991, new entrants to the profession in Texas will have to have both a disciplinary degree and no more than 18 course-hours of education courses.[43]

Currently, NSTA and NCTM both require considerable amounts of subject-specific coursework of applicants for their own certification programs. Content, rather than titles, of the courses future teachers take is essential; there is a large grey area that colleges and universities can exploit in specific subject areas (such as mathematics education). But the long-term trend is to emphasize specific skills for specific subjects rather than generic "education" courses.

Preservice education of science and mathematics teachers presents a surfeit of issues and little consensus over how to address them. College departments of science and mathematics prepare their students to become scientists or engineers, not teachers of these subjects. Few, if any, courses are offered that give prospective teachers a sense of what scientists do or how science and mathematics impact on workday activities and societal problems. Can teachers be blamed for not taking what is not offered, or for not executing in their classrooms what they were unable to experience as students (i.e., the apprenticeship role)? This "no-fault" explanation distributes the responsibility for the perceived shortcomings of the neophyte teacher.[44] It also transfers part of the burden to inservice training.

The Importance of Inservice Training

Once teachers are in place, as in any professional work force, they need periodic updating and time to consider how they could do their jobs better. At present, inservice training is also needed to remedy the inadequacies of many teachers' preservice preparation. A recent survey indicates that there has been an increase in the amount of inservice training taken during the school year, which has come at the expense of college-level

[41] Weiss, op. cit., footnote 10, table 45.

[42] The Holmes Group, *Tomorrow's Teachers* (East Lansing, MI: 1986). See also Lynn Olson, "An Overview of the Holmes Group," *Phi Delta Kappan*, April 1987, pp. 619-621. Subject-intensive preparation may be unrealistic for elementary school teachers. Just ask an elementary teacher what she teaches and the response will be "children" or "grade n"; a secondary school teacher will respond with "science" or "math." Most parents would probably take comfort that their child is being taught by someone who believes their primary allegiance and responsiblity is to children, not subjects (Shirley Malcom, American Association for the Advancement of Science, personal communication, August 1988).

[43] Lynn Olson, "Texas Teacher Educators in Turmoil Over Reform Law's 'Encroachment'," *Education Week*, vol. 7, No. 14, Dec. 9, 1987, p. 1.

[44] If scientists want to prescribe what science is worth knowing, they must be willing to collaborate with teachers in deciding what science is worth teaching. When should phenomena just be experienced and the underlying scientific principles withheld? Such a question beckons to an interdisciplinary team of scientists, teachers, child development specialists, and psychologists for answers.

course-taking on weekends and during vacations. Three-quarters of teachers now report taking inservice courses during the school year, compared with 59 percent 15 years ago.[45]

Another recent survey found that most mathematics and science teachers, at all grade levels, had spent less than 6 hours on inservice education in 1984. (See figure 3-3.) Secondary teachers had spent more time on inservice education than elementary teachers; over 10 percent of mathematics and science teachers in grades 10 to 12 had taken more than 35 hours of inservice education during the last year.[46]

A leading policy issue is who should be responsible for inservice education. As employers, school districts should be primary supporters of such education, but it is among the first budget items to be cut in periods of austerity. In practice, teachers are often expected to arrange and pay for such education themselves. While many teachers do participate, commentators suggest that there is a large pool of mathematics and science teachers who are never reached.[47]

Perhaps most lacking is a national commitment to the continuing education of science and mathematics teachers. Such education comes in many forms, including multiweek full-time summer institutes, occasional days to attend professional meetings, and provision of relevant research materials and work sessions on how to translate these into practice. In some areas, contacts between schools, school districts, scientific societies, State education agencies, and universities exist and are fruitful, but other areas are devoid of this support. Teachers need much better information than they are getting, particularly because of rapid changes in science and educational technology.[48]

The Federal Government supports inservice teacher education through both Title II of the Education for Economic Security Act program of the Department of Education and the NSF Teacher Enhancement Program. In the 1960s, NSF funded a large program of summer and other institutes, based at universities, for mathematics and science teachers. (See ch. 6.) Generally, these institutes seemed to have had positive effects, and their perceived excesses (for example, an emphasis on knowledge of science content) could be reduced were the concept to be revived. The bulk of the funds in the previous program went to colleges and universities; local school districts could now be partners in such education.

Another important Federal role could be a regional system of mathematics and science educa-

Figure 3-3.—Amount of Inservice Training Received by Science and Mathematics Teachers During 1985-86

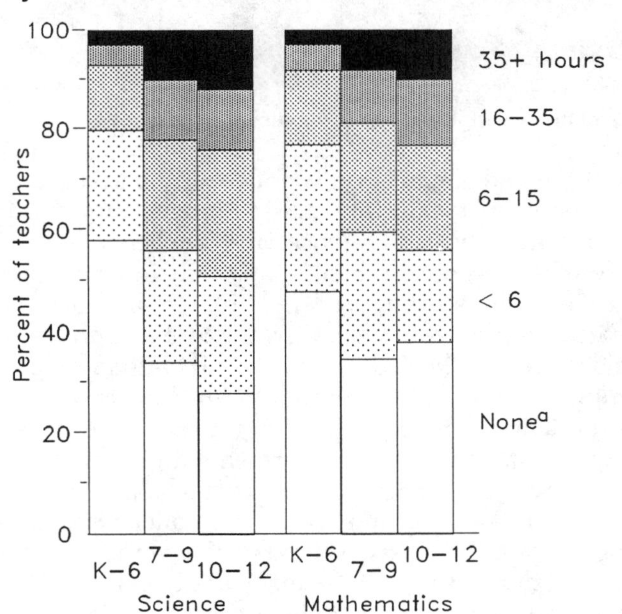

^a Includes about 10 percent with unknown time.

NOTE: Science and mathematics teachers were asked how much inservice training in science they received in the past 12 months. Inservice training includes attendance at professional meetings and workshops, but not formal courses for college credit. Sample sizes range from about 560 to 1,050, varying with grade level and field.

SOURCE: Iris R. Weiss, *Report of the 1985-86 National Survey of Science and Mathematics Education* (Research Triangle Park, NC: Research Triangle Institute, November 1987), p. 92.

[45] National Education Association, op. cit., footnote 32, tables 44-45.

[46] Weiss, op. cit., foonote 10, table 56. This difference in inservice education time may simply reflect greater opportunity afforded secondary school teachers, not lesser interest on the part of elementary teachers.

[47] This explanation raises the issue of incentives. For what does an elementary school teacher get "credit"? How do teachers perceive the relative priorities of different subject areas, e.g., language arts v. mathematics?

[48] A recent proposal is for 8 to 10 federally funded science education centers, spread around the country, which would develop curricula, train teachers, set up networks, and conduct research. See Myron J. Atkin, "Education at the National Science Foundation—Historical Perspectives, An Assessment, and A Proposed Initiative for 1989 and Beyond," testimony before the House Subcommittee on Science, Research, and Technology of the Committee on Science, Space, and Technology, Mar. 22, 1988, pp. 14-17.

Photo credit: Lawrence Hall of Science

Teacher training, important in keeping teachers up-to-date and motivated, has been a significant Federal role.

tion advisors. School administrators need training, too; they would work with school districts in disseminating the results of (at least federally sponsored) educational research, and affecting classroom practice. This role would be similar to that of county Agricultural Extension agents. The National Diffusion Network, currently restricted to conveying effective teaching curricula, is an existing mechanism for disseminating research information. Finally, the Federal Government might assist in linking teachers through informal meetings and electronic message networks. The State supervisors of science are already planning such a network.

Conclusions on Mathematics and Science Teacher Quality

The effect that good mathematics and science teaching has on students' propensity to major in science and engineering is not readily measured. Schools must lead, inform, and interest students in mathematics and science, and teachers are the front line. At the moment, many only inform and some probably dull students' interest in mathematics and science. On paper, the teaching profession is relatively well-qualified, and has had a significant (and increasing) amount of teaching experience. The teaching force needs inservice education, however; this presents an enormous task. School districts, States, and teachers (who have already had and paid for a college education) are unlikely to undertake this alone. Until school districts and States make mathematics and science teacher quality a high priority, student interest in and preparation for careers in science and engineering are not likely to flourish.

TEACHING PRACTICES AND STUDENT LEARNING

There are several teaching techniques that could be used more widely to boost both students' learning and interest in mathematics and science. In recent years, a considerable body of literature on "effective schools" has been assembled. This research has been synthesized for teachers, principals, and administrators to read.[49] There is an urgent need to write and disseminate more syntheses of this kind in other educational research areas.[50]

One technique in both mathematics and science education is experimentation. Experiments, especially when they are related to physical phenomena that students encounter in everyday life, are widely credited with improving students' attitudes toward and achievement in science. According to a recent survey, teachers think that hands-on science is an effective teaching method, yet few use it.[51] If experiments are properly planned, students learn that science advances by curiosity, manipulation, and failure. Mistakes are a normal part of science. The use of textbooks that emphasize the "facts" discovered by science, on the other hand, reinforces the popular (but mythical) view that science is a logical, linear process of accumulating knowledge.

Science experiments raise achievement scores and can often trigger positive attitudes toward science among students. Nevertheless, concerns about the cost and safety of experiments inhibit the amount of laboratory work offered, as do the limited facilities many schools have for this kind of teaching. Experiments require equipment and are more costly than lectures.[52]

Indications are that the amount of hands-on mathematics and experimental science is diminishing. (See figure 3-4.) A recent survey found that the percentage of science classes in 1985-86 using hands-on activities has fallen somewhat since 1977 at all grade levels. Hands-on activities were most common in elementary classes; only 39 percent of science classes in grades 10 to 12 used the technique (down from 53 percent in 1977). In mathematics, there have been similar declines, with the sole exception of an increase in the use of hands-on techniques in grades K-3.[53]

Other proposed teaching practices that might improve mathematics and science instruction include the use of open-ended class discussions, small group learning, and the introduction of topics concerning the social uses and implications of science and technology (often called science, technology, and society, or STS). In particular,

[49]Northwest Regional Educational Laboratory, *Effective Schooling Practices: A Research Synthesis* (Portland, OR: April 1984); and James B. Stedman, Congressional Research Service, "The Effective Schools Research: Content and Criticisms," 85-1122 EPW, unpublished manuscript, December 1985. Becoming aware of, reading about, and knowing how to apply the lessons learned, of course, are very different (Audrey Champagne, American Association for the Advancement of Science, personal communication, August 1988).

[50]In the case of mathematics and science education, the federally funded ERIC Clearinghouse for Science, Mathematics, and Environmental Education issues quarterly and annual reviews of research designed for practitioners rather than researchers. The National Association for Research in Science Teaching (NARST) also compiles summaries of current research applications in science education. The NARST series is titled *Research Matters . . . To the Science Teacher*, and is published on an occasional basis through Dr. Glenn Markle, 401 Teacher College, University of Cincinnati, Cincinnati, OH 45221. The ERIC series is a set of regular research digests in mathematics, science, and environmental education, published by the ERIC Clearinghouse for Science, Mathematics, and Environmental Education, 1200 Chambers Rd., Columbus, OH 43212. See, for example, Patricia E. Blosser, "Meta-Analysis Research on Science Education," *ERIC/SMEAC Science Education Digest*, No. 1, 1985. Finally, an ongoing project conducted by the Cosmos Corp., in collaboration with several educational associations and funded by the National Science Foundation, is collecting data on exemplary mathematics and science curriculum and teaching practices for dissemination nationally. See J. Lynne White (ed.), *Catalogue of Practices in Science and Mathematics Education* (Washington, DC: Cosmos Corp., June 1986).

[51]Eighty percent of high school science teachers agree that laboratory-based science classes are more effective than lecture-based classes, while only about 40 percent reported that they had used the technique in their most recent lesson. Weiss, op. cit., footnote 10, tables 25, 28.

[52]Indeed, experiments are a logistical nightmare for many schools: It takes teacher's valuable time to set up and tear down laboratories, assemble materials and equipment, take safety precautions, cue teacher's aides, etc. The costs—financial and otherwise—to run an experiment are often seen as prohibitive.

[53]Weiss, op. cit., footnote 10, table 25; Robert Rothman, "Hands-On Science Instruction Declining," *Education Week*, Mar. 9, 1988, p. 4. Data from the 1985-86 National Assessment of Educational Progress show that 78 percent of grade 7 students and 82 percent of grade 11 students report "never" having laboratory activities in mathematics classes. Nineteen and 15 percent of students in these grades, respectively, reported having laboratory activities either weekly or less than weekly. See John A. Dossey et al., *The Mathematics Report Card: Are We Measuring Up? Trends and Achievement Based on the 1986 National Assessment* (Princeton, NJ: Educational Testing Service, June 1988), p. 75.

Photo credit: William Mills, Montgomery County Public Schools

Hands-on science projects can be both fun and educational, and do not always require expensive equipment.

the practice of dividing classes into small, mixed-ability groups of five or six students to work on problems collectively, rather than solve them by individual competition, is widely practiced in elementary schools in Japan and is reported to be effective for students of all abilities. Its use is increasingly being advocated for the United States. The newly approved elementary mathematics curriculum in California is designed for the use of this technique, in anticipation of its wider application.[54] A recent survey found that over one-half of all students never did mathematics in small groups; only 12 percent of 3rd graders, 6 percent of 7th graders, and 7 percent of 11th graders reported using this technique daily. The survey concluded:

> Instruction in mathematics classes is characterized by teachers explaining material, working problems on the board, and having students work mathematics problems on their own. . . .
>
> Considering the prevalence of research suggesting that there may be better ways for students to learn mathematics than by listening to their teachers and then practicing what they have heard in rote fashion, the rarity of innovative instructional approaches is a matter for true concern.[55]

Because so few of these new practices are used, too many of the Nation's mathematics and science high school classes consist of teachers lec-

[54]Roger T. Johnson and David W. Johnson, "Cooperative Learning and the Achievement and Socialization Crises in Science and Mathematics Classrooms," *Students and Science Learning: Papers From the 1987 National Forum for School Science*, Audrey B. Champagne and Leslie E. Hornig (eds.) (Washington, DC: American Association for the Advancement of Science, 1987), pp. 67-94; and Robert E. Slavin, *Cooperative Learning: Student Teams* (Washington, DC: National Education Association, March 1987).

[55]Dossey et al., op. cit., foonote 53, pp. 74-76.

Figure 3-4.—Percentage of Mathematics and Science Classes Using Hands-on Teaching and Lecture, 1977 and 1985-86

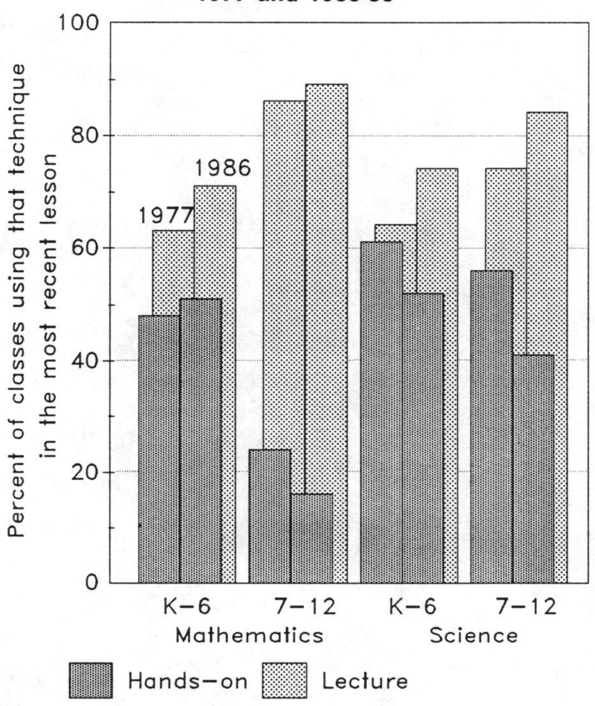

SOURCE: Iris R. Weiss, *Report of the 1985-86 National Survey of Science and Mathematics Education* (Research Triangle Park, NC: Research Triangle Institute, November 1987), p. 49.

turing about abstract material directly from textbooks. Research on teaching practices and student learning indicates that if teaching were better oriented to the way students learn, and took account of how they fit classroom knowledge into their often inaccurate world views (culled from a variety of sources), students would likely learn more and "better."[56]

Pleas for attentiveness to individual needs and learning styles possessed by different students should not be mistaken for a solution to the problems set forth in this chapter. Mathematics and science teachers are one pivotal working part in the social system known as "school."

[56]Sometimes a simple change of procedure can make a world of difference. Anne Arundel County, MD, is hoping for just such marked results, announcing its intentions of assigning the "best" teachers to students most in educational need. Will teacher assignment alone change the educational experience? Similarly, will the promotion of "master teachers" upgrade the classroom performance of teachers and students? These are school experiments intended to change the fit of the pieces in the learning puzzle.

Chapter 4
Thinking About Learning Science

Photo credit: William Mills, Montgomery County Public Schools

CONTENTS

Boxes

Table

Chapter 4

Thinking About Learning Science

We should spend less time ranking children and more time helping them identify their natural competencies and gifts and cultivate them. There are hundreds of ways to succeed, and many, many different abilities that will help you get there.

Howard Gardner, 1983

After chapters on students, schools, and teachers, it might seem odd to ask in a separate chapter: How do students learn science and mathematics? No answer will be found in the following pages. Indeed, this question now occupies several cadres of scholars and educators—neuroscientists, learning theorists, school psychologists, and an abundance of classroom teachers. Educational policy analysts have the luxury of politely raising the question, acknowledging its complexity, and substituting a slightly more tractable set of questions: How can more students be successful in science and mathematics? Does science and mathematics education search for and select a particular type of student? Is a certain learning style favored in the teaching and learning of these subjects? If so, is a self-fulfilling prophecy at work? If a certain kind of learning style is appropriate to science or mathematics, can it be promoted through programs for "gifted" or "talented" students? What can be done to correct misconceptions (held both by students and teachers), to spur creativity, to develop "higher order thinking skills," and to place more students on pathways to learning science and mathematics?[1] This chapter presents a select menu of needs, taking few orders for satisfying them.[2]

[1]Two foci of Federal support (by the National Science Foundation and the Department of Education) have been new environments for teaching mathematics and science students, and educational research on teaching and learning. Both are built on the premise that students' own experiences and intuitive explanations of scientific phenomena fuel learning. For example, see Educational Technology Center, *Making Sense of the Future* (Cambridge, MA: Harvard Graduate School of Education, January 1988); Audrey Champagne and Leslie Hornig, *Students and Science Learning* (Washington, DC: American Association for the Advancement of Science, 1987), chs. 1-2; Jan Hawkins and Roy D. Pea, "Tools for Bridging the Cultures of Everyday and Scientific Thinking," *Journal of Research in Science Teaching*, vol. 24, No. 4, 1987, pp. 291-307; Rosalind Driver, *The Pupil as Scientist?* (Philadelphia, PA: Open University Press, 1983); ERIC Clearinghouse for Science, Mathematics, and Environmental Education, *Science Misconceptions Research and Some Implications for the Teaching of Science to Elementary School Students*, Science Education Digest No. 1 (Columbus, OH: 1987); ERIC Clearinghouse for Science, Mathematics, and Environmental Education, *Secondary School Students' Comprehension of Science Concepts: Some Findings From Misconceptions Research*, Science Education Digest No. 2 (Columbus, OH: 1987); and Cornell University, Department of Education, "Proceedings of the Second International Seminar on Misconceptions and Educational Strategies in Science and Mathematics," 1987.

[2]The revision of this chapter has benefited especially from the commentary of Audrey Champagne, American Association for the Advancement of Science, personal communication, August 1988.

STUDENTS AND SCIENCE LEARNING

Students need to discover and recognize how they best learn; this aids in relating their intuitive knowledge to the knowledge conveyed in the classroom. Techniques have been devised to help children bridge their prior conceptions to the results of scientific inquiry.[3] If students' development of reasoning and analytic skills is closely

[3]For some of this pioneering work, see Joseph D. Novak and D. Bob Gowin, *Learning How To Learn* (New York, NY: Cambridge University Press, 1985).

linked to their assimilation of knowledge about particular subjects, it makes little sense to divorce the two in teaching science. Methods of inquiry and analysis, acquired via laboratory and hands-on experiences, have to be taught together with facts about particular problems or fields. All this takes time, and it may be that the amount of material covered in the typical science curriculum should be reduced, reserving instruction time for the use of hands-on techniques.[4]

The need to increase emphasis on problem solving and thinking skills is often referred to as improving students' higher order thinking skills or "creative thinking." Higher order thinking is the ability to infer and reason in an abstract way, rather than merely memorizing and recalling single items of information. These skills have always been important, but many analysts believe that they will be part of the "new basics" for tomorrow's high-technology work force.[5]

[4]In addition, computers and other interactive technologies appear to be very promising in facilitating students' construction of scientific and mathematical phenomena, and some teaching packages have been designed for this purpose. In designing effective science and mathematics education programs, experts in particular scientific fields need to pool resources with cognitive scientists and practicing teachers. See Barbara Vobejda, "A Mathematician's Research on Math Instruction," *Educational Researcher*, vol. 16, No. 9, December 1987, pp. 9-12.

[5]There is, however, no generally agreed on framework that defines the distinction between a higher order skill, a lower order skill, or any other kind of thinking skill. See Audrey B. Champagne, "Definition and Assessment of the Higher Order Cognitive Skills," *NARST Research Matters . . . To the Science Teacher* (Cincinnati, OH: University of Cincinnati, 1987); Lauren B. Resnick, *Education and Learning to Think* (Washington, DC: National Academy Press, 1987); Lynn Steen, "The Science of Patterns," *Science*, vol. 240, Apr. 29, 1988, pp. 611-616; Richard J. Murnane, Harvard University, Graduate School of Education, "U.S. Education and the Productivity of the Work Force: Looking Ahead," unpublished paper, June 1988; and *Educational Leadership*, "Teaching Thinking Throughout the Curriculum," vol. 45, April 1988, pp. 3-31.

Photo credit: William Mills, Montgomery County Public Schools

Science awards, contests, and fairs bestow public recognition on science achievers and provide hands-on experiences for many students.

The concept of higher order thinking may be a metaphor for drastic reform of schools; valid or not, many States are investing money in it.[6] A particular focus of this trend is testing, which is widely believed to be one of the main forces that perpetuates lower order thinking skills in the present day curriculum. In a sense, two related trends—a transition in the generally accepted theory of student learning and the pressure for a higher level curriculum—could lead to positive improvements in student learning of mathematics and science. But change will be slow in coming.[7]

Learning Styles

There are more similarities in how people learn than there are differences. However, there is mounting evidence that different modes (such as oral, written, and diagrammatic presentation of material) are effective with different students. These differences are believed to exist both among individuals and groups.[8]

Individual differences have been clear to teachers for years. Some students concentrate on rote memorization of facts, some explore with their hands, others are much more visual and respond best to graphical and pictorial presentation of material, while others learn via more abstract imagery. Research is beginning to clarify these differences and to explore their implications for teaching and learning.

Various models of learning styles have been devised. Each is designed to help teaching become more closely attuned to the ways different students learn, although so far these models have had limited effect in classroom practice and none is generally accepted.[9] This research builds on a long tradition of studies by philosophers, psychologists, and cognitive scientists that traces differences in learning to cultural and family backgrounds. Some differences, however, are physiological in nature, such as whether students work best at night or in the morning, and their susceptibility to extremes of light, sound, and heat. How these differences apply to the learning of science and mathematics is just beginning to be clarified. But the heavy focus of elementary and secondary mathematics and science courses on the learning and regurgitation of discrete, abstract facts has already alerted the science education community to the need for more hands-on programs, in order to break down this misleading image of what science is like.[10]

Since science and engineering have been historically populated by white males, it has been assumed that teaching approaches that are successful with this population are appropriate for all students. It is further alleged that departures from these approaches are not rewarded, and thus perpetuate the factors that deter women and most racial and ethnic minorities from considering science and engineering careers. Such deterrents are reflected in the prejudice and discrimination operating in these fields. (See box 4-A.)

As a result of this concept of how science should be taught and learned, science education in the United States, some assert, is obsessed with the testing knowledge of facts to the exclusion of four,

[6]See, for example, Rita Dunn and Shirley A. Griggs, *Learning Styles: Quiet Revolution in American Secondary Schools* (Reston, VA: National Association of Secondary School Principals, 1988).

[7]A recent initiative, called Project XL, by the U.S. Patent and Trademark Office, together with the Departments of Commerce, Energy, and Education, aims to develop inventive thinking and problem-solving skills in students. See Virginia Sowers, "Patent Office Spearheads Creative-Thinking Project," *Engineering Times*, vol. 10, No. 4, April 1988, p. 9.

[8]For example, some suggest that whites and Blacks, as well as males and females, often have characteristically different learning styles. Not surprisingly, the hypothesis that learning styles differ by race, ethnicity, and gender is highly controversial (discussed below).

[9]Christine I. Bennett, *Comprehensive Multicultural Education: Theory and Practice* (Newton, MA: Allyn & Bacon, 1986), chs. 4-5. A series of eight filmstrips produced by Madison Workshops (Dept. E-W/1379 Grace Street, Madison Schools, Mansfield, OH, 44905) are now available under the title "Learning Styles—An Alternative for Achievement." The films cover the following themes: research and learning style instruments, classification of instructional materials to accommodate student learning styles, brain dominance and learning styles, an explanation for parents, the gifted student, an approach to homework, and students with specific needs. Engineering educators are also beginning to address the importance of matching teaching and learning styles in their students. Richard M. Felder and Linda K. Silverman, "Learning and Teaching Styles in Engineering Education," *Engineering Education*, vol. 78, No. 7, April 1988, pp. 674-681.

[10]For example, see Mary Budd Rowe, "Minimizing Student Loss in Freshman and Sophomore Science Courses," *Research in College Science Teaching*, vol. 5, No. 5, May 1976, pp. 333-334; and Paul J. Kuerbis, "Learning Styles and Science Teaching," *NARST Research Matters . . . To the Science Teacher* (Cincinnati, OH: University of Cincinnati, n.d.).

Box 4-A.—A Black Learning Style?

The relation between scientific values and learning styles is reflected in the concern to improve high school completion rates among "at-risk" students. A recent booklet, written by the New York State Education Department to help teachers of at-risk students improve completion rates, included discussion of minority learning styles. A storm of protest among both minority and majority politicians and educators ensued.[1] The booklet states:

> . . .understanding various learning styles is . . . important. Several researchers have noted that the traditional classroom is built, for the most part, around the Anglo-American cultural learning styles which emphasizes the manipulations of objects such as books, listening stations, learning centers, programmed instruction and so forth. Children's racial, ethnic and emotional backgrounds and cultures influence the manner in which they learn concepts and process information. For example, qualities noted in African-Americans include:
>
> - the tendency to view things in their entirety and not in isolated parts;
> - a preference for inferential reasoning rather than deductive or inductive reasoning;
> - the tendency to approximate space, number and time instead of aiming for complete accuracy;
> - an emphasis on novelty, personal freedom, and distinctiveness; and
> - a resistance to becoming "word" dependent, but developing a proficiency in nonverbal as well as verbal communication.[2]

The existence of a Black learning style partly reflects the acceptance of Black culture in American society. At issue is whether differences arise primarily for socioeconomic or for cultural reasons, and which source is dominant. Some argue that the cultural attributes derived from membership of a particular social class are more important than cultural differences based on race. Others suggest that Blacks have created a unique Black-American culture, rooted in African traditions but adapted by the American experience, that outweighs socioeconomic factors.[3]

Several educators have argued that the cultural style and world view of Blacks had been "bastardized" by the dominant culture (the "Anglo-European cosmology") and by its educational system. White children are never asked to span and understand both Black and white cultures, let alone the subcultures within each of them, in the way that Black children are. Specifically, some claim that Black English should be accepted as a distinctive variant of standard English. The difference arises from the strong oral, rather than written, tradition of Black culture. Blacks use a more descriptive, metaphorical, context-dependent language, with few synonyms, compared with standard English. They favor the second-person ("you") rather than the third-person ("he/she"). The result is a language that is more conversational, poetic, and symbolic than that used by whites; the style in which things are said, for example in public speeches, can be as important as its content. Some argue that the idioms and concepts of Black English actually inhibit traditional mathematics and science learning. However, these same observers believe teaching can be modified and that, ultimately, Black English is as capable of expressing mathematical concepts as is standard English, although it may use different forms.[4]

A related difference occurs between the intellectual heritages and styles of inquiry of Blacks and white-Europeans. Some Blacks argue that African-American culture is more affective and "cognitively-united" in a faith-reason-emotion interdependence than is the prevailing white-European culture. Researchers on Black cognitive style speak of a "feeling intelligence" and the "aesthetic mind." They are suggesting that this style is built on what people feel and experience, and analyzes the world against that background, rather than through the European traditions of a world of universal facts and knowledge and a divorce between analysis within the mind and the feelings of the body.[5]

---

[1]Mark A. Uhlig, "Learning Style of Minorities To Be Studied," *New York Times*, Nov. 21, 1987, p. A29; William Raspberry, "Different Learning 'Styles'," *Washington Post*, Nov. 18, 1987; and Janice E. Hale, *Black Children, Their Roots, Culture, and Learning Styles* (Provo, UT: Brigham Young University Press, 1982), esp. chs. 1-2, 7.

[2]The New York State Education Department, "Increasing High School Completion Rates: A Framework for State and Local Action," a working paper of the Board of Regents, July 1987, pp. 15-16.

[3]See Kofi Lomotey, "Black Principals for Black Students: Some Preliminary Observations," *Urban Education*, vol. 22, No. 2, July 1987, pp. 173-181.

[4]Eleanor Wilson Orr, *Twice as Less: Black English and the Performance of Black Students in Mathematics and Science* (New York, NY: W.W. Norton & Co., 1987).

[5]Because commonly used teaching techniques are rooted in the European tradition, they can be ruthless with Blacks' different cultural and intellectual heritage. The educational system beats the "affective" out of students; its highest value is the neutral lecture-style teaching format, in which the lecturer shuns emotional involvement, eye contact, or voice modulation, and the students are passive absorbers of facts. See James A. Anderson, "Western Educational Systems in Conflict With Learning Styles of Minority Students," presented at the Second National Conference on Black Student Retention in Higher Education, Atlanta, GA, Nov. 4, 1986.

Not all Black educators agree on the importance, or even the existence, of a Black learning style in science and mathematics. They point out that there is a single accepted language of scientific and mathematical concepts that students who wish to progress in these fields must understand. Thus, although there might be different teaching approaches, they must all converge on the universally accepted language and content of mathematics and science.

Learning style differences may also apply to Native Americans and other racial and ethnic groups. One study of the Navajo culture suggests that, in that culture, students learn skills by watching a competent adult performing the action and then gradually taking over more of the action. Finally, the student goes off to perform the skill in private, to verify that he or she has mastered the skill. All this is accomplished with a minimum of oral communication. In schools, however, students are expected to acquire and demonstrate skills almost simultaneously, to test them in public (in front of the teacher and other students), and to learn and test skills orally. It is argued that this clash of styles of learning can seriously inhibit learning by Navajo students.[6]

Science and mathematics education that recognizes diversity will contribute more to the health of science and engineering than one of narrow gauge that alienates bright, creative, risk-taking students. Science would both benefit and change from this recognition, as the skills that the existing system culls out would refine definitions of scientific "productivity" and "creativity."

[6]Christine I. Bennett, *Comprehensive Multicultural Education: Theory and Practice* (Newton, MA: Allyn & Bacon, 1986), pp. 96-97.

and possibly other, domains: processes, creativity, attitudes, and applications.[11] In their most colorful summaries, some observers have argued that the United States, by accident rather than by design, practices "Westist, sexist, and testist" science education.[12] This approach, if accurate, is more harmful than simply deterring women and minorities from entering science, for it also discourages those whose talents lie in the other four domains and who might also contribute to science.[13]

[11]Robert E. Yager, "Assess All Five Domains of Science," *The Science Teacher*, vol. 54, No. 7, October 1987, pp. 33-37.

[12]See Howard Gardner, "Beyond the IQ: Education and Human Development," *National Forum*, vol. 68, No. 2, spring 1988, pp. 4-7, and accompanying articles in this special issue, under the title "Beyond Intelligence Testing."

[13]Evelyn Fox Keller, a scholar of women in science, has vigorously championed this point of view. She states that:

> The exclusion of values culturally relegated to the female domain has led to an effective 'masculinization' of science—to an unwitting alliance between scientific values and the ideals of masculinity embraced by our particular culture. The question that directly follows from this recognition is: To what extent has such an alliance subverted our best hopes for science, our very aspirations to objectivity and universality?

See Evelyn Fox Keller, "Women Scientists and Feminist Critics of Science," *Daedalus*, vol. 116, No. 4, fall 1987, p. 80; and Evelyn Fox Keller, "Feminism and Science," *Sex and Scientific Inquiry*, Sandra Harding and Jean F. O'Barr (eds.) (Chicago, IL: University of Chicago Press, 1987), pp. 233-246.

SCIENCE-INTENSIVE ENVIRONMENTS

One institutional response to individual and group differences in learning has been the creation of educational environments that give students greater exposure to mathematics and science than they get in regular schools and classes. Some schools specialize in these subjects. Many schools also provide special classes, including those in mathematics and science, for the so-called gifted and talented. Many students also participate in science outside the classroom, for example, in research participation programs in science laboratories. (See ch. 5.) Special schools and classes are clearly designed to have special effects on children, such as nurturing or maintaining their interest, or expediting and enriching their progress through the regular mathematics and science curriculum. This section reviews data on the extent of these special environments and the effects they have.

An Overview of State Programs and Schools

A few States have established special regional or statewide schools for mathematics and science, often in conjunction with private funding from industry. Illinois, Louisiana, Michigan, North Carolina, Pennsylvania, Texas, and Virginia directly fund statewide schools that provide specialized subject area study. (See table 4-1.) A total of 15 States sponsor, in whole or in part, schools that focus on science and mathematics; and two more are reportedly making plans to follow suit.

In addition or as an alternative to special schools, a number of States sponsor summer enrichment programs in mathematics and science. These activities are less costly than special schools and consequently more popular with the States. Twenty States offer summer programs in science and mathematics, although some of them have only very small enrollments. Florida appropriated over $1.2 million last year for such programs.

More than 30 States also have programs designed to improve the participation of women and minorities in science and mathematics. Several Northwestern States have programs designed for Native Americans.[14] Various States have begun sponsoring special recognition programs for students in mathematics and science, such as State fairs and knowledge bowls. California's Golden State Examination, established under the Hughes-Hart Educational Reform movement in 1983, is designed to identify and recognize honors-level achievement by students in specific subject areas, which include mathematics and science, and is the most comprehensive State-sponsored program nationally.

Many recognition and award programs are privately supported by professional organizations or business and industry, or jointly supported by several sources (including State and Federal Government). These types of programs often begin at the local or regional level and end at the national level; Invention Convention and Math-Counts are examples. The Westinghouse Science Talent Search and the West Virginia National Youth Science Camp are privately funded national recognition programs.

[14]Council of Chief State School Officers, *Equity and Excellence: A Dual Thrust in Mathematics and Science Education* (Washington, DC: November 1987).

Table 4-1—State-Funded Schools That Specialize in Science, Mathematics, and Engineering

Illinois
- Illinois Mathematics and Science Academy, Aurora

Louisiana
- Louisiana School for Science, Mathematics, and the Arts, Northwestern State University, Natchitoches

Michigan
- Kalamazoo Area Mathematics and Science Academy

North Carolina
- North Carolina School of Science and Mathematics, Durham

Pennsylvania
- Pennsylvania Governor's School for the Sciences, at Carnegie-Mellon University
- Pennsylvania Governor's School for Agriculture, Pennsylvania State University

Texas
- Science Academy of Austin

Virginia
- New Horizon Magnet School for Science
- Roanoke Governor's School for Science and Mathematics, Roanoke
- Thomas Jefferson High School for Science and Technology, Alexandria
- Central Virginia Governor's School for Science and Technology

SOURCE: Office of Technology Assessment, 1988; based on Education Commission of the States, *Survey of State Initiatives to Improve Science and Mathematics Education* (Denver, CO: September 1987); and data from the National Consortium of Specialized Schools in Mathematics, Science, and Technology.

Science-Intensive Schools

Science-intensive schools provide special environments for the study and practice of science and mathematics. Such schools are thought to attract students interested in science and engineering (instead of converting students to such careers), but national data are lacking to support or refute this contention. Such schools tend to attract teachers as well, and are reputed to provide high-quality instruction in mathematics and science. They are generally popular with parents, if for no other reason than they expand the choices beyond the neighborhood public school.

There are three types of science-intensive schools: well-established city-sponsored mathematics and science schools; State-sponsored schools; and magnet schools in urban areas, created to promote racial desegregation, which have mathematics or science as their theme. (See box 4-B.)

Box 4-B.—Science-Intensive Environments: Two Examples

1. The Governor's School of Science and Mathematics, Durham, North Carolina

This school became the first publicly financed residential high school in the United States devoted to science and mathematics when it opened in 1980. Funded directly by the State legislature, together with private donations, it has become a unique and exciting model for precollege science and engineering education. A recent survey found that about 80 percent of its graduates went on to science and engineering majors in college and two-thirds went to college within the State.[1]

The school is located in a former hospital in Durham, North Carolina, and is part of the educational and scientific infrastructure that fuels the continuing economic development of the Research Triangle region. The Governors' School, however, differs considerably from a regular school. It is not run by a school board, but is under the governance of the University of North Carolina system, and its teachers are exempted from certification requirements, an innovation that has attracted many who would otherwise teach only at the college level (half of the teachers have doctorates). The school enrolls 475 juniors and seniors drawn from all over the State, though enrollment is scheduled to rise to 600 by 1991, and many of the students are residential. In the first 4 years of its operation, the school received $19 million from the State and $7 million from private sources, mainly companies. Education at the school is about four times more expensive than the average for the State, costing about $10,000 per student annually. Admission to the school is on the basis of test scores, high school grades, student essays, interviews, and home school recommendations. The school's admissions committee pays particular attention to ensuring a gender, racial, ethnic, and geographic balance of its enrollment. In 1984, 47 percent of those enrolled were female and 16 percent were Black, Hispanic, or Native American.

The school stresses individual inquiry and group cooperation. Its goal is to enrich the traditional high school curriculum rather than accelerate it. The school particularly encourages students to become involved in research at nearby Duke University in Durham, the University of North Carolina in Chapel Hill, North Carolina State University in Raleigh, and with firms at Research Triangle Park.[2] More than 4,000 teachers from other schools in the State have participated in teacher training workshops held at the school. Such cooperation is valuable because the other schools in North Carolina are not nearly so lavish as the Governor's School and have not had much attention devoted to them.

2. Kansas City, Missouri, Magnet Schools

The public schools in Kansas City, Missouri, have recently announced plans to implement, by 1992, what some have described as "the most comprehensive magnet school court order in history," under which all secondary and half of all elementary schools would be designated magnet schools.[3] The Kansas City school system currently enrolls 36,000 students, of which 62 percent are Black and 6 percent are other minorities. This program of extensive magnet schools has followed long and messy court litigations that have reached the Supreme Court twice, involving the surrounding suburban school systems, the State of Missouri, and a multitude of parents along the way. The program's costs are estimated at $196 million over 6 years, part of which will be borne by Federal funds, and part by increased State and local taxation.[4]

The basis of the magnet program now being implemented is that students can follow the same theme from grade 1 to grade 12, with more choice of themes being offered at the higher grades. The themes range from the conventional ones of science and mathematics, computers, and visual and performing arts, to environmental science, engineering technology, health professions, law and public service, the military, Latin grammar, and classical Greek.

The Kansas City magnet school program is the most ambitious, and potentially most exciting, in the country. Those who will implement it face many problems, including those of funding, renovating buildings, overcoming considerable local and political suspicion, and finding enough teachers. For example, attempts have already been made to bring in teachers with the requisite skills from Belgium.

[1]Quoted in *Education Week*, June 24, 1987, p. C24. Also see Charles R. Eilber, "The North Carolina School of Science and Mathematics," *Phi Delta Kappan*, June 1987, pp. 773-777.

[2]It is not clear what relationship, if any, North Carolina Central University, a historically Black institution located in Durham, has with the School of Mathematics and Science.

[3]The following is based on Phale D. Hale and Daniel Levine, "The Most Comprehensive Magnet School Court Order in History: It's Happening in Kansas City," presented at the Fifth International Conference on Magnet Schools, Rochester, NY, May 4, 1987.

[4]Tom Mirga and William Snider, "Missouri Judge Sets Steep Tax Hikes for Desegregation Plan," *Education Week*, Sept. 23, 1987.

Some large city school systems have had intensive science and mathematics schools for many years; examples are New York City, Chicago, and Milwaukee. One of the best known is the Bronx High School of Science in New York City. These schools were founded in the early years of this century, when publicly funded high schools were a novelty, and were strongly prevocational in nature. These schools have strong national reputations for the quality of their programs and the distinguished scientists and engineers among their alumni. Each is publicly supported alongside regular schools in the same district.[15]

These schools offer a broader range of courses in mathematics and science than regular schools and normally are far better equipped. Laboratory work is a regular feature of courses, and, often, the schools have good linkages to local firms and research laboratories. These linkages help provide equipment, mentors, and, for some students, opportunities for participation in research. Teachers at these schools are often freed by school boards from many of the regular constraints on curriculum, and can work together with their colleagues to devise more coherent sequences of material than are customarily used. And, because students are generally enthusiastic and talented, teachers are keen to teach in these schools. In addition to science and mathematics, students take the other subjects they would take in a regular school.

A new organization, the National Consortium of Specialized Schools in Mathematics, Science, and Technology, was established in April 1988 to share experiences among and represent science-intensive schools. The initial meeting of the consortium included 27 schools, and more have joined since. The consortium is planning meetings for both students and faculty in science-intensive schools.[16]

Magnet schools differ from the other two types of science-intensive schools in that their primary goal is to promote racial desegregation rather than science education. Such schools, which normally have a predominantly minority enrollment, offer enhanced programs of instruction in particular areas or "themes" designed to draw students from a range of racial and ethnic backgrounds. Popular themes include emphases on particular subject areas, such as science and mathematics; on teaching methods, such as Montessori programs; or on educational outcomes, such as concentration on "the basics" or preparation for college attendance. Magnet school programs have become a popular alternative to forced busing, and have grown in number from none 20 years ago to over 1,000 today.[17]

Magnet School Issues

Since the original objective of promoting racial desegregation has largely been achieved, magnet schools are now rapidly evolving with the trend toward increased choice in public education. The concept of "schools of choice" is now an important force in education. School districts that employ magnets are realizing that all their schools never were the same; each has its own culture and interests. Rather than maintaining uniformity, the concern is to develop schools of different specialties and emphases and to capitalize on the special advantages of each school and community.[18]

---

[15]Because of the novelty of most of these schools, data on the eventual fates of their graduates are not yet available. The city science-intensive schools also have surprisingly little systematic information on their graduates. One study, conducted 20 years ago by the Bronx High School of Science, is rumored to have shown that 98 percent of its students go on to college, with less than half majoring in science or engineering. One thing is certain: only a tiny percentage of each State's high school students attend such schools. This, of course, raises questions of costs and benefits to all—students, teachers, and "regular" schools—who are part of a school system that includes a special school.

[16]The consortium is currently headquartered at the Illinois Mathematics and Science Academy, Aurora, IL.

[17]There are no current data on the number of magnet schools nationally. A 1983 survey put the number at about 1,100, of which about 25 percent had a mathematics or science theme. The magnet programs are located in more than 130 of the largest urban school districts. See Rolf K. Blank et al., *Survey of Magnet Schools: Analyzing a Model for Quality Integrated Education*, contractor report to the U.S. Department of Education (Washington, DC: James H. Lowry & Associates, September 1983). For a survey of magnet schools see Editorial Research Reports, *Magnet Schools*, vol. 1, No. 18., May 15, 1987.

[18]*Education Week*, "The Call for Choice: Competition in the Educational Marketplace," vol. 6, No. 39, June 24, 1987, supplement.

Magnet schools raise many issues. Among them are whether the introduction of such schools creates a two-class school system—magnet and nonmagnet. Magnet schools are also, on average, more expensive to run than nonmagnet schools. Another issue is the potential drain of talented students and teachers from the rest of the school system. Since local politics and school organizations vary so considerably nationwide, there are no simple rules for resolving these issues.

The school district's choice of admissions system to magnet programs is especially important. In Philadelphia, for example, the reliance placed on admissions tests by the magnet schools has been controversial because of its adverse effects on Black and Hispanic students. Alternatives to achievement test scores as the method of admission include home school recommendations, lotteries, and queues. The latter two methods are becoming increasingly popular.

Almost universally, students and parents cite the choices that they are given in magnet systems as inspiring them to have higher expectations of public education than they had before. Higher expectations should lead to better performance. There is some evidence, however, that while magnet programs improve the quality of education in schools or school districts with a high minority enrollment, minority students are sometimes less likely than white students to be admitted to magnet programs. This is because magnet programs are generally designed to draw white students into predominantly minority schools, the goal being an enrollment that better reflects the racial and ethnic composition of the school district. In some cases, limits have been put on minority enrollment so that the composition targets for the school can be met. Such limits can reduce, in the end, the access of minority students to magnet programs.[19]

From a public policy perspective, magnet schools are promising but unproven. They are designed to promote the goals of equity and excellence simultaneously, at somewhat increased cost. Anxieties about the cultivation of elites seem largely to have been diffused in most working magnet systems. Superintendents, administrators, principals, and teachers report enjoying working in magnet schools, and the magnet schools are effective mechanisms for minority advancement. The key is that magnet schools move the burden of rules, monitoring, certification, and control from administrators, school boards, and States to teachers and principals.[20] This enthusiasm, however, must be tempered by another realization: in many school districts, students do not even have the opportunity to learn science in elementary school. In addition, most schools face a serious shortage of equipment for teaching science, as well as cutbacks in resources available for offering "wet" laboratories in high schools. In short, the existence of magnet schools is no panacea to the problem of making a sequence of science and mathematics instruction accessible to more students.

Programs for Gifted and Talented Students

An increasing number of States and school districts are making special provisions for students they consider to be especially "gifted and talented." Twenty-three States now mandate such provisions, and more are considering such a policy. State spending on such provisions is rising.[21]

[19]Eugene C. Royster et al., *Magnet Schools and Desegregation: Study of the Emergency School Aid Act Magnet School Program*, contractor report to the U.S. Department of Education (Cambridge, MA: Abt Associates, Inc., July 1979).

[20]Linda M. McNeil, "Exit, Voice and Community: Magnet Teachers' Responses to Standardization," *Educational Policy*, vol. 1, No. 1, 1987, pp. 93-113.

[21]The Council of State Directors of Programs for the Gifted, "The 1987 State of the States' Gifted and Talented Education Report," mimeo, 1987, lists State programs in some detail. In addition, the Council for Exceptional Children estimates that 15 States have special certification requirements for teachers of the gifted and talented. State and local expenditures have increased and are now about $384 million (about $150 per gifted and talented child). See Council for Exceptional Children and the Association for the Gifted, testimony before the House Subcommittee on Elementary, Secondary, and Vocational Education of the Committee on Education and Labor, on H.R. 3263, The Gifted and Talented Children and Youth Act of 1985, May 6, 1986. Industry is also becoming more active, for example, through mentor programs. Gifted and talented programs are quite common in other countries where nurturing the best minds and talents is defined as a necessity. See Bruce M. Mitchell

(continued on next page)

These programs cover all subjects, including performing arts, languages, mathematics, and science. Nevertheless, the provision of programs for these students remains controversial for two reasons: the difficulty of defining criteria for identifying giftedness and talentedness, and the equity and social implications of giving these students special treatment.

While it is to be anticipated that the ranks of the gifted and talented are especially productive of future scientists and engineers, there are no data on the number of such students that eventually major in science and engineering. The Council of Exceptional Children, an advocate for gifted and talented education (as well as special education of the learning disabled and handicapped), estimates that there are about 2.5 million gifted and talented students nationally, or about 5 percent of all students. Other estimates put the number at 5 million.[22]

There is little consensus nationally on the characteristics of the gifted and talented. Standardized multiple-choice intelligence and achievement tests are widely used to sort students; those who score above threshold values on these tests are admitted to gifted and talented programs. Because of heavy reliance on such tests, there is some suggestion that labeling is prone to cultural, class, and racial bias. The process of labeling as gifted and talented does not appear to be color-blind; it has been estimated that only 13 percent of the gifted and talented are Black and Hispanic students, whereas about 25 percent of all school students are Black and Hispanic. As a result of these possible biases and of differing cutoffs and definitions of gifted and talented, the proportion of each State's school-age population labeled as gifted and talented varies considerably. One study took 18 commonly used criteria for giftedness and found that, when applied to fifth-grade suburban Minneapolis classes, 92 percent of the students could be labeled gifted in some way.[23]

The basic issue in identifying the gifted and talented is whether individuals so labeled should merely have demonstrated good progress and high achievement in schoolwork or whether they should have some truly extraordinary skills that may be undeveloped or unexpressed. Critics of gifted and talented programs suggest that most programs merely identify those who have done well in the existing system of education. In other words, the existing intelligence and achievement tests measure only a limited range of the "gifts" and "talents."

Both proponents and opponents of special provisions for the gifted and talented agree that other dimensions besides "intelligence" and "achievement" should be explored and used. Such dimensions might also include intellectual, creative, artistic, leadership, and physical and athletic abilities. Techniques for labeling need to address each of these domains separately.[24] Some States, such as Illinois and Mississippi, are making special attempts to bring students with different strengths into gifted and talented programs.

Even if there were agreement on what gifted and talented means, the provision of special programs is politically and socially contentious. Proponents of special provisions for the gifted and talented rely on a conviction that such students possess extraordinary talents not possessed by the entire population, and that these talents should be developed to the fullest possible extent. Opponents consider the creation of such programs to be elitist in nature, in practice serving the middle and upper class students almost exclusively.

Federal support for gifted and talented education programs is provided under Chapter 2 of the Education Consolidation and Improvement Act. One estimate suggests, however, that only about 20 percent of school districts use any of their

(continued from previous page)

and William G. Williams, "Education of the Gifted and Talented in the World Community," *Phi Delta Kappan*, March 1987, pp. 531-534.

[22]Definitions of "gifted," "talented," and "special" students tend to fluctuate with the annual amounts of Federal and State funding available in these categories. This is also why a demographic bulge in a particular grade will disqualify students from "GT" (gifted and talented) classes that they took in the previous grade. The criteria are arbitrary, but their interpretation (e.g., all those in the nth percentile and above) is often rigid.

[23]Lauren A. Sosniak, "Gifted Education Boondoggles: A Few Bad Apples or a Rotten Bushel," *Phi Delta Kappan*, vol. 68, March 1987, pp. 535-538.

[24]Robert J. Sternberg and Janet E. Davidson (eds.), *Conceptions of Giftedness* (Cambridge, England: Cambridge University Press, 1986); and Howard Gardner, "Developing the Spectrum of Human Intelligence," *Harvard Educational Review*, vol. 57, No. 2, 1987, pp. 187-193.

Chapter 2 funds for gifted and talented education.[25] Several bills introduced in the 100th Congress were designed to fund model programs for educating the gifted and talented, training the teachers of such students, and expanding research on gifted education.

The Department of Education's Office of the Gifted and Talented was disbanded in 1981. There are calls for its reinstatement to ". . . carefully coordinate the use of limited federal resources and to provide a much needed focal point of national leadership."[26] The recent reauthorization of Federal education programs includes a provision requiring the Department of Education to set up a National Center for Research and Development in the Education of Gifted and Talented Children and Youth.[27] Center proponents argue that Federal support has a catalytic role vis-a-vis States and school districts, and that the current reform movement has neglected the gifted and talented in favor of the mainstream and learning disabled.

The Council for Exceptional Children estimates that only one-half of gifted and talented students receive any kind of special assistance, and that such assistance itself is limited so that most gifted and talented students still spend substantial portions of time mainstreamed in ordinary classes. The Council estimates that about $400 million nationally is spent each year on such special assistance, but only about $10 million of Chapter 2 funds are spent on such programs. Some support also comes from Title II of the Education for Economic Security Act. Some of this funding has been spent on science-intensive schools, such as the North Carolina School of Science and Mathematics, which is sometimes included under the rubric of "gifted" education. Proponents argue that this funding, even though it is increasing, is still too little. There is, however, increasing interest within the private sector in such programs. In addition, several university-based programs, such as those at Johns Hopkins, Ohio State, and Duke, identify talented individuals, including those in mathematics and science, and provide enrichment programs for them during the summer.

Given that gifted and talented children have been identified and that special provision will be made for them, the basic educational issue is this: whether gifted and talented programs should focus on enriching students' exposure to the existing curriculum or encouraging them to accelerate their progress through that program so that they complete the traditional sequences of high school courses a year or two early. A related debate concerns whether students should receive enrichment or accelerated classes in all subject areas or only in single subjects, such as mathematics. A final issue is whether such focused instruction should be provided in dedicated "special" schools, or as an adjunct to the regular school curriculum.

For able students stifled in the conventional, slow-moving educational system, gifted and talented classes can provide relief and progress suited to their intellectual and emotional needs. Such classes can also help keep such students in school; many of those who drop out of school are bored, but gifted, children. The basic argument for special treatment of the gifted and talented is that without it these students would be ignored or unchallenged by the existing school system.

[25]Ellen Flax, "Economic Concerns Aiding Programs for Gifted," *Education Week*, vol. 6, No. 33, May 13, 1987, pp. 1, 17.

[26]Ibid. Also see The Council for Exceptional Children and the Association for the Gifted, testimony before the Senate Subcommittee on Education, Arts, and Humanities of the Committee on Labor and Human Resources, on Reauthorization of Chapter 2 of the Education Consolidation and Improvement Act, July 16, 1987, p. 7.

[27]Public Law 100-297, "Conference Report to Accompany H.R. 5," Report 100-567, April 1988, p. 115.

CLOSING THOUGHTS: A LARGER MENU?

The debate over gifted and talented students begs very different questions about educational methods than the debate over alternative learning styles. The problem addressed by educators concerning different learning modes is the negative reinforcement and frustration many otherwise talented people experience in the traditional classroom. In mathematics and science learning, this has tended especially to be the case with women, and racial and ethnic minorities. The teacher of gifted and talented students faces a different problem. These students have already demonstrated some proficiency or success in the present system; the educator's task is to sustain student interest and progress.

Both issues have similar implications for science and mathematics education: How can more students be successful in science and mathematics? What does it mean to be talented in science or mathematics? How should such talent first be identified, and then nurtured? Are special schools or programs needed? Will innovative curricula that reflect new insights into how students learn —and how different their learning styles may be—spark the interest and fulfill the potential that teachers and parents often recognize in their children? Will new thinking about learning penetrate the schools? Will it be effective in "calling" more students to science and mathematics, helping them fulfill expectations (rather than ill-founded prophecies), while propelling them to the next educational stage and, ultimately, a career in science and engineering?

These questions reflect high expectations for science and mathematics education. Indeed, this chapter has glimpsed a larger menu of issues, needs, and signs of progress.

Chapter 5

Learning Outside of School

Photo credit: Don Mills, Ontario Science Center

CONTENTS

Boxes

Table

Chapter 5
Learning Outside of School

The children who see colored shadows on the wall in the Exploratorium have not "learned" something you can test. But years later, when a teacher tries to explain light patterns and perception, this experience will be a part of the deep background that will make learning easier.

George W. Tressel, 1988

National concern about the excellence of American schools has sometimes detracted from the recognition that children learn a great deal out of school. The influences of family, friends, the media, and other features of the environment outside of school are profound.

The out-of-school environment offers opportunities to raise students' interest in and awareness of science and mathematics. Table 5-1 presents estimates of the proportion of the school-age population that participate in science-related informal education activities. Such informal activities draw strength from the local community—churches, businesses, voluntary organizations, and their leaders. All are potential agents of change. All are potential filters of the images of science and scientists transmitted by television and other media. These images are often negative—nearly always intimidating—and shape young people's views of science as a career.

Science centers and museums, for example, can awaken or reinforce interest, without raising the spectre of failure for those who lack confidence in their abilities. (As Frank Oppenheimer, founder of San Francisco's famed Exploratorium, noted, "Nobody flunks a museum.") Intervention programs, aimed especially at enriching the mathematics and science preparation of females, Blacks, Hispanics, and other minorities can rebuild confidence and interest, tapping pools of talent that are now underdeveloped.

Table 5-1.—Estimated Proportions of Target Populations That Participate in Informal Science and Engineering Education Programs

Activity	Proportion
Occasional viewers of *3-2-1 Contact*	—50 percent of 4- to 12-year-olds
Regular viewers of *3-2-1 Contact*	—30-35 percent of 4- to 12-year-olds
Did a science-related activity after viewing *3-2-1 Contact*	—25 percent of 4- to 12-year-olds
Visit a science center or museum	—25 percent of school-age students each year
Visit a science center or museum	—50 percent of 4- to 12-year-olds
Visit an aquarium or zoo	—90 percent of 4- to 12-year-olds
Take an inservice course at a science center or museum	—Less than 1 percent of teachers
Participate in an intervention program	—Less than 1 percent of Black and Hispanic students
Participate in an intervention program	—Less than 0.1 percent of female students
Enroll in a weekend or summer science enrichment program	—0.1 percent of high school graduates

SOURCE: Office of Technology Assessment, 1988.

THE IMPORTANCE OF FAMILIES

Family circumstances are pivotal influences on career choice. Parents' occupations, attitudes, incomes, residences, and socioeconomic class are all reflected in their children's lives. Much of a child's initial learning about the world, and about reading, speaking, and writing, takes place in the family. Families can give children a head start in preparing for school, in progressing through the educational system, and shaping perceptions of careers.[1]

[1]This is the motivation for the American Association for the Advancement of Science's LINKAGES Project, and particularly for the appeal for minority parental involvement in their children's education, as illustrated by The College Board, *Get Into the Equation:*

(continued on next page)

Although it can be illustrated in many ways, the strong influence of families is indicated by biographic data on winners of the Westinghouse Science Talent Search (WSTS). The WSTS is one of America's oldest high school competitions in the sciences. Between 1942 and 1985, it awarded $2 million to 1,760 young scientists. Its winners have gone on to earn five Nobel Prizes, two Fields Medals, and four MacArthur Foundation ("genius") Awards. Two surveys of previous winners, one conducted in 1961 and another in 1985,[2] suggest that parents, close relatives, or teachers played critical roles in their decisions to become scientists. Male family members were especially important influences (the bulk of the winners were male). For example, in the 1985 survey, 35 percent of winners had fathers who were professional scientists or engineers. Among other influences, 62 percent of the WSTS sample cited a professor or a teacher as playing a major role in their career decision, and 44 percent reported that they became interested in their current professional fields in high school.

Parental involvement in education has always been recognized as important. An innovative mathematics program called Family Math, based at the Lawrence Hall of Science in Berkeley, California, is designed to encourage parents to work

(continued from previous page)
Math and Science, Parents and Children (New York, NY: College Entrance Examination Board, September 1987).

[2]Harold A. Edgerton, *Science Talent: Its Early Identification and Continuing Development* (Washington, DC: Science Service, Inc., 1961); and Science Service, Inc., *Survey of Westinghouse Science Talent Search Winners* (Washington, DC: Westinghouse Electric Corp., November 1985).

Photo credit: William Mills, Montgomery County Public Schools

Parents are instrumental in shaping their children's attitudes toward education.

with children in solving problems and learning mathematics. Having parents learn something about science and mathematics will, in turn, help children learn.[3]

[3]David Holdzkom and Pamela B. Lutz, *Research Within Reach: Science Education; A Research Guided Response to the Concerns of Educators* (Charleston, WV: Appalachia Educational Laboratory, 1984), pp. 192-202; Beth D. Sattes, Educational Services Office, Appalachia Educational Laboratory, "Parent Involvement: A Review of the Literature," unpublished manuscript, November 1985; Jean Sealey, Appalachia Educational Laboratory, "Parent Support and Involvement," *R&D Interpretation Service Bulletin in Science*, n.d.; Jean Kerr Stenmark et al., *Family Math* (Berkeley, CA: Regents of the University of California, 1986); and The College Board, op. cit., footnote 1.

THE PUBLIC IMAGE OF SCIENCE

The public image of science and engineering conveyed by television and the other media is ambiguous. On the one hand, science is portrayed as being of great benefit to economic progress and to health. On the other, it is portrayed as a sinister force that bestows power on its adherents and is manipulated by inhumane people.[4] In any case, the process of scientific and technological advance is poorly understood by the public. While the sometimes dismal image of science is one part of the cacophony of discouraging signals that an aspiring young scientist receives (and will certainly cause some students to shun science), evidence suggests that poor images of science are probably not a leading cause of students' failing to pursue careers in these fields. Academic preparation ultimately is far more important.[5]

Television Images of Science

The public image of science has been studied in a number of separate settings over the years. One study dissected the content of network prime-time dramatic programs between 1973 and 1983.[6] The study found that: 1) some aspect of science and technology appears in 7 of every 10 dramatic television programs; 2) doctors are more positively portrayed than are scientists; and 3) scientists are not as successful in their on-screen occupations as other occupational groups. In fact, for every scientist in a major role who fails, two succeed, whereas for every doctor who fails, five succeed, and for every law enforcer who fails, eight succeed. The study also documented a link between these images and viewers' attitudes and concluded that, while television dramas generally presented positive images of science, the more that viewers see, the more they perceive scientists as odd and peculiar.

Television is a pervasive influence on many students' lives.[7] It is argued that the effects of television viewing may be strongest for children under 11 years old, because up until that age children are continuing to develop interpretive skills and sophistication in analyzing the content of material.[8] Children's attitudes toward televised material will be influenced, in other words, by what they have seen and learned of the world. As children grow older, television is critical in the formation of an understanding of social relationships and of the social forces that govern adult life. This window on the adult world may be particularly important in shaping attitudes toward careers and occupations, because adolescents have few social contacts outside their own age group.[9]

Television viewing can have positive and negative effects. It can promote racial and sexual stereotypes and perceptions of occupational segregation, or help change attitudes toward the races

[4]See, for example, Spencer Weart, "The Physicist as Mad Scientist," *Physics Today*, June 1988, pp. 28-37.

[5]The following is based on Robert Fullilove, "Images of Science: Factors Affecting the Choice of Science as a Career," OTA contractor report, September 1987.

[6]George Gerbner, "Science on Television: How It Affects Public Conceptions," *Issues in Science and Technology*, vol. 3, No. 3, winter 1987, pp. 109-115.

[7]Unpublished data from the mathematics and science assessments of the 1986 National Assessment of Educational Progress indicate that 24 percent of Blacks in grade 11 and 44 percent of Blacks in grade 7 watch 6 hours or more of television daily, compared to 9 percent and 24 percent for the whole populations in grades 7 and 11, respectively (Marion G. Epstein, Educational Testing Service, personal communication, June 1987).

[8]Fullilove, op. cit., footnote 5, pp. 30-36.

[9]Gary W. Peterson and David F. Peters, "Adolescents' Construction of Social Reality: The Impact of Television and Peers," *Youth and Society*, vol. 15, No. 1, September 1983, pp. 65-85. Also see Joan Ganz Cooney, "We Need a 'Sesame Street' for Big Kids: Television Can Help Our Children Learn Math and Science for the '90s," *Washington Post*, Sept. 11, 1988, pp. 16-17.

and sexes, depending on the content of programming.[10] Heavy television viewing reduces the time students spend on homework, thus depressing their academic performance. However, television viewing by adolescents is reported to have remained roughly constant over the time that, for example, Scholastic Aptitude Test scores have fallen.[11]

From a policy perspective, however, even if a link between television portrayals of science and engineering and career aspirations were established, the challenge would be to design television programming that could affect those aspirations positively. Research suggests that it is easier to change attitudes, and therefore aspirations, than to change behaviors.[12]

Students' Images of Science

In a nationwide study sponsored by the American Association for the Advancement of Science (AAAS) in 1957, high school students were asked to compose short essays describing their impressions of scientists and their work. Margaret Mead and Rhoda Metraux wrote a composite description of science and of scientists from their reading of 35,000 such essays.[13] They found that:

> The number of ways in which the image of the scientist contains extremes which appear to be contradictory—too much contact with money or too little; being bald or bearded; confined to work indoors, or traveling far away; talking all the time in a boring way, or never talking at all—all represent deviations from the accepted way of life, from being a normal friendly human being, who lives like other people and gets along with other people.

A 1977 study, 20 years after the work of Mead and Metraux, found that perceptions had changed little. Over 4,000 children from kindergarten to grade five in Montreal, Canada, were asked to draw pictures of what they thought a scientist looked like. The dominant image of scientists found by Mead and Metraux a generation earlier was held by younger students as well. In addition, more elements of the stereotype appear as students advance through the grades.[14]

The Potential of Educational Television

Educational television can be a powerful way to introduce new images and teach students. A prominent example is the Children's Television Workshop's *3-2-1 Contact*, funded by the National Science Foundation (NSF) and the Department of Education, and broadcast daily on most public television stations. This show's target audience is children 8 to 12 years old (although many younger children watch as well). Its aim is to interest these students in science, with particular emphasis on female and minority children. Extensive research has been done on this series. It is estimated that the series is seen in nearly one-quarter of all households with at least one child under 11 years old. Themes in *3-2-1 Contact* are echoed in series-related science clubs and a maga-

[10] Fullilove, op. cit., footnote 5.

[11] Mark Fetler, "Television and Reading Achievement: A Secondary Analysis of Data From the 1979-80 National Assessment of Educational Progress," presented at the Annual Meeting of the American Educational Research Association, April 1983; and Barbara Ward et al., *The Relationship of Students' Academic Achievement to Television Watching, Leisure Time Reading and Homework* (Washington, DC: National Institute of Education, September 1983). Television viewing has been cited as one of the possible causes for the decline in the Scholastic Aptitude Test scores of America's college-bound students that occurred in the 1970s. See College Entrance Examination Board, *On Further Examination: Report of the Advisory Panel on the Scholastic Aptitude Test Score Decline* (New York, NY: 1977). See also, U.S. Congress, Congressional Budget Office, *Educational Achievement: Explanations and Implications of Recent Trends* (Washington, DC: U.S. Government Printing Office, August 1987), pp. 69-71.

[12] Icek Ajzen and Martin Fishbein, *Understanding Attitudes and Predicting Social Behavior* (Englewood Cliffs, NJ: Prentice Hall, 1980); and J. Baggaley, "From Ronald Reagan to Smoking Cessation: The Analysis of Media Impact," *New Directions in Education and Training Technology*, B.S. Alloway and G.M. Mills (eds.) (New York, NY: Nicholds Publishing Co., 1985).

[13] Margaret Mead and Rhoda Metraux, "Image of the Scientist Among High-School Students," *Science*, vol. 126, Aug. 30, 1957, pp. 384-390.

[14] David Wade Chambers, "Stereotypic Images of the Scientist: The Draw-A-Scientist Test," *Science Education*, vol. 67, No. 2, 1983, pp. 255-265. The stereotypes apparently persist into adulthood, although for many citizens the ambivalence toward science never subsides. Etzioni and Nunn found in a review of national public opinion polls on the attitudes of Americans toward science that most Americans value science for its contribution to the Nation's high standard of living; similarly, Americans hold generally favorable opinions of scientists and trust their judgment. But images of what a scientist does remain fuzzy, and opinions on science vary significantly by age, education, region, socioeconomic class, and personality type. Amitai Etzioni and Clyde Nunn, "The Public Appreciation of Science in Contemporary America," *Science and Its Public: The Changing Relationship*, G. Holton and W.A. Blanpied (eds.) (Boston, MA: D. Reidel, 1977), pp. 229-243.

Photo credit: William Mills, Montgomery County Public Schools

Television can be a powerful adjunct to printed materials and classroom instruction. Video technologies in general can let children see science in action.

zine, and it is estimated that one-half of all viewers have done some science-related activity.[15]

The impact of a television message may depend more on the characteristics of the viewer, such as his or her age, than on the characteristics of the message. OTA finds no compelling support for the hypothesis that poor images of science by themselves deter students from science and engineering careers. Television is a powerful force, both for good and for bad, but its effect also depends on the prior experience and knowledge of viewers.

[15]Research Communications, Ltd., "An Exploratory Study of *3-2-1 Contact* Viewership," National Science Foundation contractor report, June 1987.

The Informal Environments Offered by Science Centers and Science Museums

Science centers and museums aim to bring science alive with exhibits and displays that show scientific phenomena in action, and dedicated staffs determined to spark interest in science. They have important positive effects on students' attitudes toward science and knowledge of physical phenomena.

Science museums were first set up to house and archive the achievements of science and technology through artifacts such as experimental apparatus, machines, field notes, and pictures. As attitudes toward science and science learning have changed, new institutions, called science centers, have sprung up with the primary aim of exciting and educating visitors in science and technology rather than of chronicling its history.[16] Indeed, this development was presaged by science museums. At the turn of the century, the Deutsches Museum in Munich, West Germany, was the first science museum to invite the public to participate in its exhibits, and to introduce cutaways and working models to encourage the public to learn about how things work rather than what they look like. This model was soon copied for use in the new Chicago Museum of Science and Industry in the 1930s. The preeminent example of a science center is the Exploratorium, in San Francisco, which opened in 1972.

Today, both science museums and science centers use "hands-on" exhibits to illustrate scientific principles through "object-based learning" that schools and school systems often cannot provide because the equipment is too expensive or unavailable.[17] Science centers are also making increasing use of new technology such as computers, video, and videodiscs, both as exhibits in their own right, and as a means of illustrating scientific and technological concepts.

[16]See Sheila Grinell, "Science Centers Come of Age," *Issues in Science & Technology*, vol. 4, No. 3, spring 1988, pp. 70-75.

[17]See Michael Templeton, "The Science Museum: Object Lessons in Informal Education," *NSTA Annual 1987*, Marvin Druger (ed.) (Washington, DC: National Science Teachers Association, forthcoming).

Learning in Science Centers

The learning that occurs in a science center is grounded in the relevance of science to practical daily life, in arousing curiosity, and in allowing people to explore and explain for themselves. Most important, it has no particular aim; no tests are in sight, no goals are prescribed or proscribed, and visitors are free to choose the exhibits that they would like to explore in detail and to ignore the rest. The function of these centers, from the point of view of the education of future scientists and engineers, is not so much to teach scientific concepts, as to interest students and to relate abstract learning in schools to experience of how physical phenomena work and can be altered by humans.

Nationally, there are about 150 centers and museums dedicated to improving public understanding of science and technology; an umbrella organization, the Association of Science-Technology Centers (ASTC), represents most of these centers. A recent survey by ASTC indicated that these 150 centers had about 45 million visitors in 1986, up from 32.5 million in 1979.[18] Part of this increase is due to expansion in the number of science centers; over one-half opened after 1960, and 16 percent opened after 1980. The survey suggests that as many children and young people as adults visit the centers. Probably about 6 million children and young people come on school trips. Interest in science centers is very high. Once largely the preserve of large cities, centers are now being built in many smaller towns, cities, and rural areas.

The people who work in science centers are skilled at helping both adults and children learn about science. Science centers' target audience is the population that is curious but not confident about science. About 100 science centers conduct mathematics and science teacher training programs, funded by school districts, States, or Title II funds from the Federal Government. These programs often aim at elementary and middle school teachers, many of whom have almost no grounding in science. About 65,000 teachers participated in such programs in 1985. Few centers have programs for high school teachers.[19]

Photo credit: Nancy Rodger, Exploratorium

Science centers, which allow children to touch and play with equipment and exhibits, expose them to scientific concepts in an appealing setting.

Science centers have developed close working relations with school systems in the areas in which they serve, while remaining independent of them. This unique role is often cited as being useful, because it allows the centers to take risks and experiment in science education in ways that school systems find difficult. Nancy Kreinberg, Director of the EQUALS program (see box 5-A) based at the Lawrence Hall of Science in Berkeley, California, has put it this way:

> We are in the schools, but we are not of the schools. We are in the community, but we don't represent one faction of the community. We are seen as representing a lot of different interests, and I think that is an enormous source of strength that every science center has to offer.[20]

A novel program run by the San Diego School District sends every fifth grader in the city to spend a week at Balboa Park, the city's museum district. A special team of teachers spends the week with the students, exploring both art and science museums. Although designed primarily to assist racial desegregation (the students are formed

[18]These and data cited below are from Association of Science-Technology Centers, "Basic Science Center Data Survey 1988," unpublished.

[19]Jacalyn Bedworth (ed.), *Science Teacher Education at Museums: A Resource Guide* (Washington, DC: Association of Science-Technology Centers, 1985).

[20]Association of Science-Technology Centers, *Natural Partners: How Science Centers and Community Groups Can Team Up to Increase Science Literacy* (Washington, DC: July 1987).

Box 5-A.—The EQUALS Program

EQUALS is a program designed to improve the awareness of gender- and race-related issues in mathematics and science education. It encompasses projects for teachers, counselors, administrators, parents, and school board members to promote the participation of female and minority students in mathematics and computer courses. EQUALS provides curriculum materials, staff development seminars, and family learning opportunities. It is located at the Lawrence Hall of Science (University of California at Berkeley). Since 1977, 14,000 California educators and 9,000 educators from 36 other States and abroad have participated in EQUALS courses. Sites have been established across the United States.

Evaluation data from extensive interviews and questionnaires indicate that EQUALS programs increase student enrollment in advanced science and mathematics classes, improve attitudes and interest in related occupations, enhance the professional growth of teachers, and, perhaps most importantly, encourage parent involvement in the schools (the Family Math Program is currently being evaluated with National Science Foundation funds). EQUALS publications, including *We All Count in Family Math* and *I'm Madly in Love With Electricity and Other Comments About Their Work by Women in Science and Engineering*, are geared for use in all types of "classrooms," since the emphasis is on hands-on, active problem solving. Materials in Spanish are also available and widely used.

SOURCE: EQUALS Program, Lawrence Hall of Science, University of California at Berkeley.

into heterogeneous groups of 5 to 10 students drawn from different schools, neighborhoods, races, and ethnicities), the program makes use of existing facilities that are neither available to schools nor would readily fit into existing school learning patterns and curricula.[21]

Costs and Benefits

A recent survey indicates that the average science center costs about $1.5 million per year to run, and that most charge between 75 cents and $5 for admission.[22] About 40 percent of expenses incurred by the average center are defrayed by admissions, memberships and other fees, and from sales of souvenirs and food; State and local districts pay, on average, about 28 percent of the costs, while corporations contribute another 10 percent. The average contribution of the Federal Government to ASTC member centers is 6 percent, but the bulk of this goes to the three centers it wholly supports.[23] The remaining centers receive, on average, just 2 percent of their income from Federal sources.

Several Federal programs fund science centers, including the Informal Science Education Program of NSF, the National Endowment for the Humanities, the National Endowment for the Arts, the Institute of Museum Services, and the Department of Education's Secretary's Discretionary Fund. Only the Institute of Museum Services will contribute toward routine operating expenses (and it sets a limit on its contribution of $75,000 per museum per year); the other sources will fund only particular programs and novel educational projects. Indirect Federal support has in the past also come from contributions of equipment and facilities. (Seattle's successful Pacific Science Center, for example, is housed in the United States pavilion built for the 1962 World's Fair, which was given free to the center.)

Evaluations of the effects of science centers on students are limited. The research that has been done indicates that science centers can be effective arenas to demonstrate aspects of the natural world, but have more limited impacts in conveying understanding of the scientific concepts underlying particular exhibits. Visitors often acquire lasting memories of phenomena, such as the formation of a rainbow by the use of a prism, but are less readily able to explain what they have seen or give the proper scientific terms that describe the phenomena. Written information beside exhibits is not often well assimilated. Visitors thus

[21]Judy Diamond, Natural History Museum, San Diego, personal communication, June 1988.

[22]All data in this paragraph are from Association of Science-Technology Centers, op. cit., footnote 18. Note that this database excludes a few science centers and museums that are not association members.

[23]These three centers are the Air and Space Museum and the National Museum of American History in Washington, DC (both part of the Smithsonian Institution), and the Bradbury Science Museum, Los Alamos National Laboratory, New Mexico.

build up a good intuition of how things work, based on their experience of phenomena, but little analytical knowledge. Students who are used to figuring things out for themselves, ignoring instructions, often find science centers interesting; the style of learning that science centers employ is radically different from that in formal classrooms where the emphasis is often on obeying rules and memorizing facts.

Learning that takes place in science centers is thus difficult to measure using conventional tests of factual recall (which do not demonstrate "learning" at all), but is clearly important. When families visit and explore exhibits together, parents can often become more confident about science and hence more supportive of any interest that their children might develop.[24] An interesting study of the "Explainer" program at the Exploratorium, in which nonscientifically inclined but enthusiastic high school students explain particular exhibits to visitors, found that, 10 years later, former Explainers were still very interested and confident in science, academic pursuits, and work experiences. (See box 5-B.)

ASTC is working to improve attendance and use of science centers by females and minorities, and is encouraging its members to form links with community and service organizations in the female and minority communities, such as the National Urban League, Girls Clubs of America, and the National Action Council for Minorities in Engineering.[25] Several foundations are helping fund such outreach programs. Minority students, in particular, often need to be encouraged to develop interests in science and engineering, and science centers can help build their confidence in these areas. Several science centers have held highly successful "camp-ins," in which students or teachers spend a whole night learning and playing in a science center.

Informal Learning

Informal education, then, is not just museums and science centers. Informal learning also takes place through reading, watching television, visiting libraries, and participating in clubs. It is this additional informal education, as one NSF staffer puts it—4-H Clubs, Girl's Club of America, Girl Scouts—that warrants ". . . a concerted effort to give kids direct hands-on experience."[26]

[24]Diamond, op. cit., footnote 21.

[25]Association of Science-Technology Centers, op cit., footnote 18. The American Association for the Advancement of Science's Office of Opportunities in Science, through its LINKAGES project, has been the source of many activities spearheaded by the Association of Science-Technology Centers.

[26]George W. Tressel, "The Role of Informal Learning in Science Education," presented to the Chicago Academy of Sciences, Nov. 14, 1987, p. 11.

INTERVENTION AND ENRICHMENT PROGRAMS

Some kinds of informal education programs are designed to enrich, or even replace, traditional schooling in mathematics and science. One form is the "intervention program," designed to improve educational opportunities for special groups not often well served in regular classrooms (particularly females and minorities). Other programs for the entire school population allow students to participate, for example, in science experiments in research laboratories, including Federal laboratories, or to enhance their progress through the regular school mathematics and science curriculum. These programs are known as enrichment progams.

Intervention Programs

Ideally, all students would have access in school to high-quality courses in mathematics and science, and their teachers, fellow students, and guidance counselors would be sensitive to the overt and covert racism and sexism that interferes with learning. In practice, however, the quality of courses is very uneven, and social attitudes still deter females and minorities from pursuing further science and engineering study. While schools are reforming and improving the situation, change is slow and certainly lagging the demographic changes already occurring. Negative attitudes

Box 5-B.—Learning by Teaching: The Explainer Program at San Francisco's Exploratorium

One of the most powerful ways to learn is by teaching others. The Explainer program at San Francisco's Exploratorium, a science center designed to offer visitors maximum involvement with scientific phenomena and experiments, gives a small group of enthusiastic teenagers just this chance. Explainers are located around some of the Exploratorium's exhibits to help visitors by conducting demonstrations, answering questions, and sparking discussion about the concepts that the exhibits convey. The Explainer program is intended both as a service to visitors to the Exploratorium, and as a work and educational program to give teenagers an appreciation for science and for learning.

The Exploratorium recruits Explainers from local high schools for their enthusiasm for working with the public, rather than for their interest in science as such. The program deliberately reaches out to students outside the academic and science mainstream; good grades or interest in science are not prerequisites, and in fact are not desired. Potential Explainers have to be friendly and keen to help visitors, and represent diversity among the population. Over one-half of Explainers are from minority groups; they are equally divided between males and females.

Explainers are hired for a 4-month session, are paid an hourly wage, and receive about 50 hours of paid training before and during the job. The program costs the Exploratorium about $250,000 per year (or about $4,000 to $5,000 per Explainer). Most Explainers work only one session.

An evaluation of nearly 900 alumni of the Explainer program was conducted in 1985-86.[1] Former Explainers universally report that their stint at the Exploratorium was a tremendous learning and social experience, as well as a boost to their self-esteem. One of the greatest benefits the Explainers cited was working intensely with a small, diverse group (15 to 20 Explainers work with visitors at any one session), and enjoying professional cameraderie with the Exploratorium staff. Explainers acquired confidence in their ability to learn about subjects they had previously thought inaccessible. They learned to deal with not knowing "all the answers"; they also developed communication and people skills that they later found valuable at college and in the workplace.

Among the comments made by former Explainers were these:

> There would be times when something didn't catch my interest in class, but it did when I learned it here. It was hands on. There was actual proof. It wasn't something read from a textbook.
>
> I learned to tolerate a lot of my own mistakes. . . . You learn to appreciate that you can learn from those that know better. Once at an eye dissection, I got into a conversation with an ophthamology student. I'd be explaining things but all of a sudden I was learning new stuff by talking to this guy.
>
> It got rid of a stigma for me and let me go and pursue science, which is really what I wanted to do in the first place. I found out that, yeah, you can enjoy science and you're not weird if you do, so why not? Before, I would just keep it to myself. I never told anybody that I read science books before I came here.
>
> That was one of the key things to come out of the Exploratorium experience: becoming a people-oriented person. When you explain something and you see the spark in people's eyes, you are enriching them. You are giving them something, and in return you're getting the feeling that you are enriching their lives.
>
> Part of the reason I liked it a lot was that it gave me the feeling that I was teaching for the first time. I was showing people things instead of always having them shown to me.

In sum, the curiosity and desire to learn that Explainers acquired stayed with them in their later lives.

Women were much more likely than the men to report that they became interested in science and engineering and improved their communication skills as a result of the Explainer program. Students who were already interested in science and engineering strengthened their confidence; other students gained general self-esteem and were encouraged to go to college.

The Explainer program also helps visitors enjoy themselves and learn. Explainers can particularly help reach their peers—other teenagers who traditionally have been tough customers for science centers and

[1]Judy Diamond et al., The Exploratorium, "The Exploratorium Explainer Program: The Long-Term Impact on Teenagers of Teaching Science to the Public and a Survey of Science Museum Programs for High School Students," mimeo, June 1986. The Explainer program has operated since the opening of the museum in 1969. For the study, 32 representative alumni were interviewed at length, and a questionnaire was developed on the basis of those interviews and sent out to former Explainers. Other information on Explainers was gathered from interviews with museum visitors and applicants to the Explainer program.

museums to reach. Compared to other age groups, it is believed that comparatively few teenagers attend museums; one study found that many teenagers felt that most museum staff were "patronizing and condescending."

Some other science centers have adopted the Exploratorium model directly and many others have developed other outreach programs for teenagers, including summer classes, mentorships, interpretive guides, and museum staff internships.[2] Many target high-ability, science-oriented students; others are directed to enthusiastic students who are willing to take the job on a volunteer basis.

[2]Ibid.

toward careers in science and engineering conveyed by influences outside schools, such as the media, families, and friends, will not readily be altered.

Against this background, special efforts to intervene must be made in order to attract and encourage females and minorities in science and engineering. Most of these efforts, though few in number, take place out of school during students' free time. Although intervention programs have only been in existence for about 20 years, many successful techniques have been developed for boosting the self-image, enthusiasm, and academic preparation of females and minorities for science and engineering careers. Indeed, some of these techniques (such as stressing the relevance of science understanding to everyday experiences, the use of small groups, and participation in hands-on activities) clearly warrant dissemination to the entire population of students.

The Content and Reach of Intervention Programs

The Office of Opportunities in Science of AAAS collects data on intervention programs, and is an enthusiastic advocate of them.[27] The programs differ from each other considerably in terms of their longevity, bases of operation, sources of support, goals, and quality. Universities, museums, and research centers house the majority of intervention programs, and many serve junior high and high school students. Most effective intervention programs involve learning science by doing, rather than through lectures or reading; working closely with small groups of other students; contact with attentive advisors, mentors, and role models who foster self-confidence and high aspirations; and an emphasis on disseminating information about science and engineering careers. Intervention programs for minority students often reach a high percentage of females as well, both minority and majority. Evaluations suggest that early, sustained intervention can bring minority achievement to the same level as that of white males.

AAAS has examined exemplary intervention programs and has found that they have strong leadership, highly committed and trained teachers, parental support, adequate resources, a sustained focus on careers in science and engineering, clear goals, and continual evaluation. Many combine academic and informal learning, and involve teachers and parents. They often focus on enriching students' experiences in science, rather than in providing remedial treatment for the poor quality experiences that most students have had from formal education; many also stress techniques, such as peer learning, that help students learn how to learn. The intervention programs that work best start early in students' educational careers and have a long-term focus, with the ultimate goal of making successful intervention techniques part of the normal apparatus of the school system.

Most intervention programs require extraordinary staff commitment and support, and are not easy to replicate in other locations. The most talented teachers and leaders can only fully serve a limited number of students, even using technol-

[27]American Association for the Advancement of Science, Office of Opportunities in Science, "Partial List of Precollege Mathematics and Science Programs for Minority and/or Female Students by State," unpublished manuscript, July 1987; and Shirley M. Malcom et al., *Equity and Excellence: Compatible Goals; An Assessment of Programs That Facilitate Increased Access and Achievement of Females and Minorities in K-12 Mathematics and Science Education*, AAAS 84-14 (Washington, DC: American Association for the Advancement of Science, Office of Opportunities in Science, December 1984).

ogies such as distance learning. Some programs have, however, been replicated. Perhaps the most successful is the Mathematics, Engineering, and Science Achievement (MESA) program based in Berkeley, California. MESA-modeled programs now operate in about 10 other States.

Differences Among Intervention Programs

Intervention programs recognize the different problems that can face females and minorities in science and engineering. Most females, for example, have access to the high school science and mathematics courses that they would need for science and engineering careers; the issue is one of self-image and self-confidence. Some female students mistakenly believe they do not even need to take optional mathematics and science courses for entry to science and engineering majors in college. In addition, during group work in classrooms, male students frequently dominate experimental equipment and computers, leaving female students taking notes and acting as "secretaries." In all-female intervention programs, each student can fully participate in operating equipment and enjoy the whole experience of making scientific observations. Intervention programs need to focus on encouraging an interest in science and improving student self-confidence in mathematics and science.

Many minorities, on the other hand, do not have access to the necessary mathematics and science courses and are less likely than whites to plan to attend college. While many Blacks and Hispanics are interested and aware of science and engineering careers—historically a route to social mobility—they lack the preparation to enter them. Accordingly, intervention programs need to improve the probability that minorities will be prepared to attend college at all, and then focus on improving their learning of mathematics and science.[28]

Within the minority population, however, there are significant differences that affect the design of intervention programs. Many Mexican-Americans come from poor rural backgrounds and have strong family bonds, but tend to receive little encouragement at home for "book learning." Cuban-American students often come from well-educated families and do very well in academic coursework. Black students in northern cities may be aware of the rewards of science and engineering, but are often poor and enrolled in poorly funded school systems; their access to necessary courses is limited. Black students in the South, however, are more likely to live in rural areas, and have less knowledge of (and correspondingly, interest in) science and engineering careers. Programs for Black students are needed early in their educational careers, because deficiencies in preparation accumulate at an early age. Asian-Americans are often very well prepared for science and engineering careers, but those from territories of the United States with significant Asian populations, Pacific Islanders such as American Samoa, tend to lack preparation. Boxes 5-C, 5-D, and 5-E illustrate a variety of intervention programs.

Funding Intervention Programs

Despite the effort that has been put into developing intervention programs in the last two decades, and the urgent need for them, there is still only a modest number of programs and, collectively, they reach only a small proportion of their target populations. The leaders of several major intervention programs meet as the National Association of Precollege Directors (NAPD), which estimates that intervention programs reach 40,000 minority students annually (or less than 1 percent of the total minority student population). But 25,000 participants in NAPD programs have graduated from high school, over half to major in science or engineering.[29] Expansion is limited both by the shortage of individuals prepared to commit the time and energy necessary to initiate these programs and by lack of funding. A local base of support seems to be an essential ingredient of success.

Some intervention programs owe their origins to Federal funding. Many today are supported by States, foundations, and industry. Federal funds,

[28]For discussion of the more general goal, see Gloria De Necochea, "Expanding the Hispanic College Pool: Pre-College Strategies That Work," *Change*, May/June 1988, pp. 61-65.

[29]Joel B. Aranson, "NSF Initiatives—A Minority View," *Opportunities for Strategic Investment in K-12 Science Education: Options for the National Science Foundation*, Michael S. Knapp et al. (eds.) (Menlo Park, CA: SRI International, June 1987), vol. 2, p. 112.

Box 5-C.—National Council of La Raza

The National Council of La Raza is headquartered in Washington, DC, and has program offices in Phoenix, Edinburg (Texas), and Los Angeles. Over the last several years, the organization has developed and demonstrated five innovative community-based approaches to improve the educational status of Hispanics. Three of the five projects cater to precollege students in special "at-risk" populations; the other two focus on the needs of parents and teachers. Original support was provided by the American Can Co. Foundation, and further funding has come from AT&T and Carnegie Corp. grants. Projects in Kansas City, Phoenix, and Houston recently received grants from Time, Inc. and the Xerox Corp. Projects in several other cities have community and foundation funding.

The council's educational programs supplement school offerings, their rationale being that enrichment programs improve the educational experience of Hispanic children more than do remedial programs that repeat school lessons. The council serves as a national advocacy organization to encourage systemic reforms in teacher training, continuing education, and effective school practices. Project coordinators are confident that much change can be initiated through community-generated local projects. The five community-based approaches are the Academia del Pueblo, Project Success, Project Second Chance, Parents as Partners, and the Teacher Support Network.

The problems of early academic failure and the large number of Hispanic children who must repeat grades are addressed by the Academia del Pueblo, which provides after-school and summer "academies" for elementary school-aged children. These efforts help students meet and exceed grade promotion requirements. Project Success provides career and academic counseling to help junior high students raise their expectations and to support their eventual progress to high school graduation. Project Second Chance targets dropouts using volunteer mentors and tutors. The Parents as Partners program was designed to reinforce the concept that parents are effective teachers, a concept that is particularly influential in Hispanic communities. This project trains and assists parents to encourage and be tutors to their children. The Teacher Support Network brings together community resources to train and support both Hispanic and non-Hispanic teachers who work with Hispanic children.

The council assists demonstration sites with the necessary training and technical assistance to implement the models, and monitors and evaluates the projects. The council also transfers lessons arising from demonstration projects. A necessary component of any council program is the development of participants' Spanish language skills, either as an integral component of the curriculum or as a second language. In addition, the council is assisting the Association of Science-Technology Centers to identify Hispanic community-based organizations with a mathematics and science education focus to encourage their participation in science centers and museums.

however, play an important base role. Since many intervention programs piggy-back on existing facilities in schools, science centers, research laboratories, and universities, the programs are relatively inexpensive. They are labor- not capital-intensive, and many have budgets of several hundred dollars per student. (College-level programs tend to be more expensive, sometimes around several thousand dollars per student, although this sum might include tuition and scholarship support.)

Intervention programs were one outgrowth of the civil rights movement of the 1960s. Federal law eventually was extended to address many forms of discrimination, by race, sex, handicap, and age; intervention programs began to attract Federal funding as a way of breaking down some of the barriers to full participation of these groups in science and engineering. One source of the funding was the Women's Educational Equity Act of 1974, administered by the Department of Education. Another was grants for State programs, funded under Title IX of the Higher Education Amendments of 1974. Federal funds have often acted, and been most effective, as seed funding to initiate intervention programs; if successful, such programs have sometimes then been funded by States and industry in their community.

By contrast, NSF historically has not emphasized either intervention programs or other ways

Photo credit: William Mills, Montgomery County Public Schools

Intervention programs, like the summer research programs for Hispanic students shown here, give students a chance to work together on real research projects.

Box 5-D.—American Indian Science and Engineering Society

The American Indian Science and Engineering Society (AISES) is a nonpolitical, national organization of American Indian scientists and engineers. The society's primary purpose is to expand American Indian participation in science and engineering careers and to promote technical awareness among other Indians. Since it was founded in 1977, AISES has grown to represent 61 tribes in 36 States and Canada. Projects are designed to encourage academic excellence at all education levels. The precollege programs coordinate teacher training seminars and summer enrichment programs, provide materials (including computers), introduce role models through science fairs and camps, sponsor competitions, coordinate student chapters, and publish newsletters and videotapes. Collegiate programs include mentorships, internships, and workshops. The Collegiate Chapter Program includes an annual 2-day conference for leadership training and focuses on providing scholarship information and peer support at 35 institutions. Professional programs are also available for Indian scientists and engineers.

Funding sources are public and private, including the National Science Foundation, the Department of Education Fund for the Improvement of Post-Secondary Education, the National Aeronautics and Space Administration, Hewlett Packard, and Hughes Aircraft. The society also earns some income from the publication of *Winds of Change*, a quarterly magazine designed to disseminate information on educational opportunities and AISES activities, and to promote involvement of Indian and non-Indian participants in Indian concerns. To benefit from AISES programs, schools and agencies must affiliate with the society. The society's annual conference brings together Indian students, the affiliates, and non-Indian professionals from the public and private sectors.

SOURCE: American Indian Science and Engineering Society, Boulder, CO.

Box 5-E.—Philadelphia Regional Introduction for Minorities to Engineering

Founded in 1973, Philadelphia Regional Introduction for Minorities to Engineering (PRIME) is a nonprofit corporation devoted to creating opportunities for minorities in engineering, pharmacy, and other mathematics- and science-based professions. PRIME was incorporated in 1975 as a partnership of Philadelphia area businesses, government agencies, colleges and universities, professional associations, and parent groups, and caters to junior and senior high school students in the Philadelphia, southern New Jersey, and Greater Delaware Valley region. The program is funded by member organizations—including 32 companies, 6 State and local agencies, and every major local college and university in the region. Each of these organization's active interest in improving the number and quality of young minority engineers has helped ensure the program's success. Original support came from the Sloan Foundation, but institutional support has been steady. The program's success is evident in its longevity, continued expansion, and the high percentage of the program's students who graduate college and enter their chosen professions.

PRIME's identification of students interested in mathematics and science subjects begins in the seventh grade. Early intervention, it is believed, contributes to the program's success. "Capable" minority students (as defined by test scores, teacher recommendations, and scholastic achievement) are provided with specialized supplementary activities, counseling, and monitoring throughout the remainder of their academic years. Hands-on projects are emphasized in classes, and industry and government representatives visit classrooms monthly, providing role models.

Students also interact with member agencies through workshops and field trips. During the summer, PRIME offers competitive university-based residential training programs (PRIME Universities Program), sponsored originally by a grant of $50,000 from General Electric. While in college and after graduation, students receive job referral and placement assistance. Guidance is continuous; students receive advice on everything from how to cope with precollege coursework to selecting an appropriate university and obtaining financial aid. Enrollment in the summer program is 80 percent Black and 5 percent Hispanic; the academic year program includes only Black students. Both programs have equal numbers of males and females.

PRIME serves more than 3,000 students each year. Recent statistics show that 92 percent of PRIME high school graduates enter college, and 73 percent of those enter technical and/or engineering programs. Eighty-five percent earn baccalaureate degrees.

SOURCE: PRIME, Inc., Philadelphia, PA.

of improving the "career chances" of females and minorities in science and engineering. NSF programs that *have* had some success include: the Women in Science program (1974-76, 1979-81); the Resource Centers for Science and Engineering program, aimed at minorities (1978-81); and the Research Apprenticeship for Minority High School Students (1980-82). None of these programs was reestablished when NSF's Science and Engineering Education (SEE) Directorate was reborn in 1983, although the Research Apprenticeship for Minority High School Students and the Resource Centers program have recently been resurrected. With these two exceptions, none of SEE's current programs directly address "underrepresented" groups. Programs for these groups have not been well funded through other NSF efforts, although some receive funding (for example, through Science and Mathematics Education Networks, and Teacher Enhancement). SEE encourages the submission of proposals for projects to address underrepresented groups.[30]

Funding sources for intervention programs have varied according to the program's target populations. Mission agencies have set different priorities. NSF and the Department of Education have established intervention programs for females; foundations, the military, and other Federal agencies (such as the Department of Health and Human Services, the National Aeronautics and Space Administration, and the Department of Energy)

[30]While the National Science Foundation proposed two specific minority programs for fiscal year 1988 and is planning more, their approach has been criticized as insufficient for the magnitude of the problem. See ibid., pp. 111-112.

have tended to fund programs for minorities. Overall, however, there have been few sustained funders of "women in science" programs. Much targeted human resource support is done *institutionally* to categories such as historically Black colleges and universities. But this support, too, seems modest relative to the number of lives they are supposed to change and careers they claim to launch.[31]

Initially, intervention programs were based outside schools. Schools, in fact, have often been viewed by advocates of intervention programs as part of the problem rather than part of the solution. But in recent years, schools have increasingly begun to work with intervention programs such as the METRO Achievement Program in Chicago. Interventions such as the Ford Foundation's Urban Mathematics Collaboratives work directly with mathematics teachers outside the school systems in the cities that the program serves; this frees teachers from the organizational and attitudinal constraints that such systems engender. It is important that intervention programs complement efforts to improve the formal education system.

[31]OTA research found that benefactors scale down their gifts to fit their expectations of success when historically Black institutions are involved. Foundations that give $1 million to an Ivy League school will give an institution with primarily minority enrollment $100,000 for the same activity. Also see Tom Junod, "Are Black Colleges Necessary?" *Atlanta*, vol. 27, October 1987, pp. 78-81.

Enrichment Programs

Many programs are designed to enrich or speed the progress of talented individuals through science and mathematics courses. Several Federal laboratories, most prominently those of the Department of Energy, provide summer research participation programs that allow students to experience science in the flesh by active participation in real research. While there is no conclusive evidence that such programs change students' career destinations, they have a potent effect in confirming student inclinations that research can be fun. Several universities operate summer courses in mathematics and science for talented individuals. The best known are the Center for the Advancement of Talented Youth at The Johns Hopkins University and the Talent Identification Program at Duke University. (See box 5-F.)

CONCLUSIONS

It is clear that intervention and enrichment programs are a valuable supplement to formal education. The expectations and attitudes of parents and teachers; differential access to courses, instrumentation, and educational technologies; and lack of information or mentoring by role models lead many female and minority students away from science and engineering careers. The magnitude and complexity of the problem requires a large and continuing effort.

Experience with intervention programs has provided considerable knowledge on elements of successful models to replicate in future programs. The reasons why intervention programs have failed, not surprisingly, have tended not to be well documented.

When provided with early, excellent, and sustained instruction and guidance, the achievement levels of females and minorities in science and engineering can match those of any other student. In other words, there are no inherent barriers to participation. The Federal role in intervention programs is to encourage new starts, possibly to expand funding, and to provide networks for the elements of successful programs to be disseminated and shared. Some programs should be based in schools, while others should not. Tailoring each to the needs of specific populations, circumstances, and problems, is, in this domain as well as in many other areas of education, the key to success.

Box 5-F.—Duke University Talent Identification Program

The Duke Talent Identification Program (TIP) was founded in 1980 by Robert Sawyer, Associate Professor of Education at Duke University, to identify academically talented adolescents and accelerate their education.[1] The program provides them with information about their abilities and academic options, and sponsors special educational programs for them. The "Talent Search" operates in 16 States, but the program is open to all qualified students nationwide.

TIP operates four educational programs for talented adolescents: a Summer Residential Program, a Precollege Program (also residential and held in the summer, but offering college credit for coursework), a By-Mail Program, and a Commuter Program. The programs are largely financed by student fees, but financial aid is readily available. Students must take the Scholastic Aptitude Test (SAT) in the seventh grade to qualify for admission. Twelve percent of the students that TIP identified in 1986 scored higher than the average college-bound senior in mathematics, while 13 percent scored higher than average in verbal skills. A variety of academic programs are then offered to the students based on the test scores.

An average class profile is as follows: for the 3-week summer biology course, students range from 7th to 10th grade, have previously passed algebra II, and have SAT scores of 1,000 or higher (this score is generally the minimum for the seventh graders; the scores are progressively higher as the students get older). The student body averages 50 percent female, 23 percent Asian, and 15 percent other racial and ethnic minorities. Most of the students have already skipped at least one grade in their formal schooling. Evaluation consists of the College Board Advanced Placement topical essay examination and an appropriate achievement test, in addition to laboratory work and research projects. The students are not graded (except in college courses), but they are given information on their results to let them know how they are doing. The programs place emphasis on science writing and using challenging textbooks. The program has collected some followup data; it has found that 20 percent of the students go back to regular courses during the subsequent school year, and 35 percent to advanced placement courses.

TIP maintains an automated database that tracks the 140,000-plus participants, and research is in progress on the post-secondary activities of these youngsters. The program publishes a quarterly newsletter, *Insights*, that is distributed nationwide. *Insights* features reports on TIP activities, summaries of significant research on education, special program descriptions, announcements of academic competitions, book reviews, teacher and student resources, and articles on college and career options.

[1]Funding was initiated with a $280,000, 5-year grant from the Duke University Endowment; 91 percent of the program budget at present is provided by student fees, with the remaining 9 percent funded through gifts, grants, and the sale of materials to outside parties.

Chapter 6
Improving School Mathematics and Science Education

Photo credit: William Mills, Montgomery County Public Schools

CONTENTS

Boxes

Figure

Tables

Chapter 6

Improving School Mathematics and Science Education

Now when I say "education," I'm going beyond what is in NSF's Science and Engineering Education Directorate and looking at the education capabilities and programs of the foundation [NSF]. We have the responsibility to advance capabilities along the whole educational pipeline. I don't think any other agency, whether State or Federal, has that mission.

Erich Bloch, 1988

The American system of public schooling is large, diverse, and stolid. Pressure to reform various features of the system, especially curricula, graduation standards, and the education of teachers, has been building since at least the early-1980s, and some change, much of it led by the States, has been realized. The "education reform movement," as it is called, has drawn strength and encouragement from leaders in government, education, business, and higher education.[1] Much of the pressure for reform has been bolstered by economic arguments, stressing the need for international competitiveness and the industrial advantages of a well-educated work force.

[1]See, for example, National Commission on Excellence in Education, *A Nation at Risk* (Washington, DC: U.S. Government Printing Office, 1983); and Gerald Holton (ed.), " 'A Nation at Risk' Revisited," *The Advancement of Science, and Its Burdens* (Cambridge, England: Cambridge University Press, 1986).

THE SYSTEMIC NATURE OF THE PROBLEMS FACING AMERICAN SCHOOLS

The problems that face mathematics and science education in the schools are complicated and interrelated. In the broadest sense of the term, they are systemic. Reforms of any one aspect of mathematics and science teaching, such as course-taking, tracking, testing, and the use of laboratories and technology, can and have been undertaken. But each change is constrained by other aspects of the system, such as teacher training and remuneration, curriculum decisions, community concerns and opinions, and the requirements and influences of higher education.[2] Very little analysis has been undertaken of the costs and benefits of different kinds of improvements that could be made in mathematics and science education. A recent review suggested that schools':

> . . . influence on learning does not depend on any particular educational practice, on how they test or assign homework or evaluate teaching, but rather on their organization as a whole, on their goals, leadership, followership, and climate. . . . These organizational qualities that we consider to be the essential ingredients of an effective school —such things as academically focused objectives,

[2]Iris R. Weiss, OTA workshop summary, September 1987; F. James Rutherford, "Activities in Precollege Education," *Competition for Human Resources in Science and Engineering in the 1990s*, Symposium Proceedings (Washington, DC: Commission on Professionals in Science and Technology, Oct. 11-12, 1987), pp. 60-65; Arthur G. Powell et al., *The Shopping Mall High School: Winners and Losers in the Educational Marketplace* (Boston, MA: Houghton Mifflin, 1985); and Ernest L. Boyer, *High School: A Report on Secondary Education in America* (New York, NY: Harper & Row, 1983).

pedagogically strong principals, relatively autonomous teachers, and collegial staff relations—do not flourish without the willingness of superintendents, school boards, and other outside authorities to delegate meaningful control over school policy, personnel, and practice to the school itself. Efforts to improve the performance of schools without changing the way they are organized or the controls they respond to will therefore probably meet with no more than modest success; they are even more likely to be undone.[3]

Incremental and Radical Reforms

Against this background, a case can clearly be made for "starting all over" with a new system of organizing, administering, and even funding schools. The education system has evolved incrementally, and, during the last 200 years, has adapted to changing societal expectations, expanded its reach to almost the entire population of students up to age 18, been influenced by the changing economy of the United States, and responded to judicial intervention in many aspects of its organization, including its financing. These changes have been made on the superstructure of existing culture and practices, and have not necessarily resulted in the "best" system. But starting all over is not practical or politically feasible: too much is invested in the current system of mathematics and science education. The best short-term focus, therefore, will be on incremental improvements within the existing system.[4]

In 1985, Congress asked the National Science Foundation (NSF) to commission a special study of investment options for NSF to undertake in science and mathematics education. The contractor, SRI International, was asked to identify specific niches that the Science and Engineering Education Directorate of NSF could fulfill, given the existing structure, experience, and expertise of the Agency. The report of this study, published in June 1987, makes many concrete suggestions of ways NSF could use its special experience and expertise.[5] Ten current areas for future NSF investment are listed in box 6-A. The SRI report also discussed the trade-off between incremental change and wholesale renovation, arguing that while incremental improvements may not be the way to effect fundamental change, far-reaching innovations in science education not grounded in the current elementary and secondary school system will simply not be adopted.[6]

For now, incremental reform is the likely way American mathematics and science education will be improved. The remainder of this chapter examines some improvements that are taking place and others to be contemplated, against the background of intersecting local, State, and Federal interests.

Local and State Initiatives

Local and State initiatives could go a long way toward improving elementary and secondary mathematics and science education. Much can be and is being done to improve mathematics and science education at the local level of the school board and the school, from introducing magnet programs to re-equipping science facilities.

The many different initiatives spawned by schools and school districts are difficult to summarize because they do not form part of a single State, regional, or national plan. This does not detract from their importance; they can be highly beneficial. In 1983, the American Association of School Administrators (AASA) sent a questionnaire to 1,500 school administrators, mainly superintendents. From these and other data, AASA compiled a list of the top 10 most common actions already being taken by school districts to improve mathematics and science education. (See table 6-1.)

[3]John E. Chubb, "Why the Current Wave of School Reform Will Fail," *Public Interest*, No. 90, winter 1988, pp. 28-49. Also see Peter T. Butterfield, "Competitiveness Plank Seven—Education: The Foundation for Competitiveness," *Making America More Competitive* (Washington, DC: The Heritage Foundation, 1987), pp. 69-76.

[4]For the perspective of the former Secretary of Education, see William J. Bennett, *American Education: Making It Work* (Washington, DC: U.S. Department of Education, April 1988), esp. pp. 23, 31, 35, 41, 45.

[5]Michael S. Knapp et al., *Opportunities for Strategic Investment in K-12 Science Education: Options for the National Science Foundation, Summary Report* (Menlo Park, CA: SRI International, June 1987). Also see Robert Rothman, "NSF Urged to Assert Itself in Push to Improve Education," *Education Week*, Sept. 9, 1987, p. 12.

[6]Knapp et al., op. cit., footnote 5, pp. 36-39. This chapter draws on the SRI report in discussing various National Science Foundation elementary and secondary mathematics and science education efforts.

Box 6-A. —Opportunities for Future Investment in K-12 Mathematics and Science Education: Recommendations by SRI International to NSF's Science and Engineering Education Directorate

Opportunities to Devising Appropriate Content and Approach

- **Redesign and improve existing mathematics curricula at all grade levels.** The amount of repetitive computation should be reduced, and the amount of effort devoted to other topics, such as the skills of mathematical problem solving, probability and statistics, and computer sciences, should be expanded.
- **Redesign the way in which elementary school science is taught.** Elementary school science, despite NSF's attempts in the 1960s to develop "hands-on" curricula, is still very limited in scale and depth. NSF has begun an initiative in this area.[1] **Similarly, one priority should be to redesign and improve middle and high school curricula.**
- **More effort should be made to match mathematics and science education with the needs and backgrounds of students, particularly females and minorities.** The reach of existing programs and curricula must be extended, but experiments are also needed to tailor teaching to the special needs of each type of learner.

Opportunities to Strengthen the Professional Community

- **The people who assist mathematics and science teachers,** such as lead teachers, curriculum specialists, and science and mathematics coordinators, **need more help and support.** Multiyear training programs, recognition programs, and development of stronger alliances between higher education and school districts would enhance this support function.
- **The number training to become mathematics and science teachers needs to be increased, and their training improved.** NSF could enhance the "professionalization" of the teaching force, the content of teacher training courses, and the utilization of knowledge about teacher recruitment and training programs.
- **Strengthen the informal science education community.** Educators out of schools—on television and in museums and science centers—are becoming increasingly important. These people need training and professional development, and would benefit from larger networks and closer collaboration.

Opportunities to Leverage Key Points in Educational Infrastructure

- **Improve and expand publishing capabilities in mathematics and science education.** An emphasis on broadening the base of learners will require new and different teaching materials: current materials are largely aimed at the "science- and engineering-bound." Collaborative programs with existing publishers would help improve the textbook publishing process, and promotion of alternative publishing routes would help provide a diversity of materials that the current mechanisms of market operation are not able to support.
- **Improve testing and assessment methods and practices in mathematics and science.** The growing power of testing over curricular and teaching decisions indicates the urgency of developing and implementing tests that measure a broader range of skills, concepts, and attitudes than current fact-oriented tests. NSF's skill at research and development gives it special expertise in managing research programs in testing.
- **Work with State mathematics and science education reform leaders.** Interest among these leaders in improving the teaching of these subjects is strong, but their familiarity with the educational issues involved is limited. NSF could assist State-level groups to devise and implement reforms, and to develop networks.
- **Expand the proven power of informal education programs and assist their assimilation into schools.** These programs are effective at reaching and motivating large and diverse groups of students. Innovations are needed, as is better outreach to more communities.

[1]The National Science Foundation's first solicitation in this area, in fiscal year 1986, addressed elementary mathematics curricula, and resulted in six awards. A second solicitation, also in fiscal year 1986, addressed elementary science curricula and aimed to develop ". . . partnerships among publishers, school systems and scientists/science educators for the purpose of providing a number of competitive, high quality, alternative science programs for use in typical American elementary schools." Among these latter awards has been the Technical Education Research Center's project in linking computers and science learning. A third round of awards was made in May 1988. See National Science Foundation, Directorate for Science and Engineering Education, "Summary of Grants, FY 1984-86: Instructional Materials Development Program," NSF 86-85, unpublished document, March 1987; National Science Foundation, Program Solication, "Programs for Elementary School Science Instruction II," NSF 87-13, unpublished document, 1987; *Science*, "NSF Announces Plans for Elementary Science," vol. 235, Feb. 6, 1987, p. 630; and "N.S.F. Gives $7.2 Million for 'Hands-On' Science Material," *Education Week*, May 25, 1988, p. 19. On elementary science curricula generally, see Marcia Reecer, "Pointing Out and Disseminating," *Science and Children*, vol. 24, No. 4, January 1987, pp. 16-18, 158-160.

SOURCE: Office of Technology Assessment, 1988, based on Michael S. Knapp et al., *Opportunities for Strategic Investment in K-12 Science Education: Options for the National Science Foundation*, Summary Report (Menlo Park, CA: SRI International, June 1987), pp. 10-16.

Photo credit: William Mills, Montgomery County Public Schools

Education policy begins at the local level.

States are playing an increasing role in K-12 education, through finance, curriculum and graduation requirements, and assessment and monitoring. Most States are funding a growing proportion of the cost of public elementary and secondary education (see figure 2-1 in ch. 2), spurred by the warnings contained in the rash of educational reform reports of the early 1980s.[7] This activity has been chronicled in recent surveys by the Education Commission of the States and the Council of Chief State School Officers (CCSSO).[8]

These two reports document the ways in which States have become more active in four areas:

- curriculum requirements;
- assessment of the extent to which curriculum

[7]National Governors' Association, *Results in Education—1987: The Governors' 1991 Report on Education* (Washington, DC: 1987), pp. 36-37. "State," as used here includes the District of Columbia, Puerto Rico, the U.S. Virgin Islands, and Guam.

[8]Education Commission of the States, *Survey of State Initiatives to Improve Science and Mathematics Education* (Denver, CO: September 1987); Jane Armstrong et al., "Executive Summary: The Impacts of State Policies on Improving Science Curriculum," prepared for the Education Commission of the States, unpublished manuscript, June 1988; and Rolf Blank and Pamela Espenshade, *State Education Policies Related to Science and Mathematics* (Washington, DC: Council of Chief State School Officers, State Education Assessment Center Science and Mathematics Indicators Project, November 1987). For State actions in relation to all subjects, see ibid.; and Denis P. Doyle and Terry W. Hartle, "Leadership in Education: Governors, Legislators, and Teachers," *Phi Delta Kappan*, September 1985, pp. 21-27.

Table 6-1.—Summary of Kinds of Local Initiatives to Reform K-12 Mathematics and Science Education

The 10 most common and frequent actions being taken by school districts:

1. **Revise, reconstruct, and strengthen the science and mathematics curricula.** Committees are at work discarding old content, adding new units, and expanding the scope and sequence in established and new offerings. A major aim is to bring about an articulation of offerings, in kindergarten through grade 12.
2. **Generate new activities to retrain, reeducate, and lend a helping hand to classroom practitioners**. Inservice education, in doses more massive than ever before, goes on at an ever-increasing pace within school systems and on college campuses. Cooperation with colleges and universities is at a high level on behalf of both science and mathematics.
3. **Modernize and expand facilities** needed for science, providing better-equipped laboratories for upper grades, and offering teachers suitable working space for elementary hands-on science activities.
4. **Make available new textbooks and other instructional materials for science and mathematics.** They buy "packaged programs" (STAMM, COMP, SCIS, ESS), but, above all, districts develop their own curriculum guides, teacher resource handbooks, and units for students—all geared to local district philosophy, aims, and objectives.
5. **Raise requirements for the study of science and mathematics,** often under the spur of State legislation, at times by decision of boards of education. The big push is toward more years of science and mathematics at the secondary level, and more time spent on task in the elementary grades.
6. **Monitor science and mathematics programs more closely than ever before**. They assess, evaluate, and measure. Methodology, content, and student achievement are under close scrutiny at all times by principals, but more often by specialized personnel using new tools and instruments.
7. **Go into partnerships with industry, higher education, and community groups**. Out of these cooperative efforts—also called alliances and consortiums—come advanced content (from scientists and mathematicians); new opportunities for inservice education (from colleges and universities); and greater support for science and mathematics programs (from community and civic groups.)
8. **Devise new programs to attract and hold students** who have so far been largely bypassed by science and mathematics education—Blacks, Hispanics, American Indians, and other minorities.
9. **Support, with greater interest than ever before, extracurricular activities for science and mathematics students.** They seek the establishment of clubs and encourage greater student participation in science and mathematics fairs, olympiads, and other competitions—both for the able and the average student.
10. **Seize the role of advocacy,** sensing this is the time and opportunity to rebuild and strengthen the science and mathematics curriculums.

SOURCE: The material in this table is from Ben Brodinsky, *Improving Math and Science Education* (Arlington, VA: American Association of School Administrators, 1985), pp. 29-30.

requirements are met;

- providing special programs for female, minority, gifted and talented, handicapped, and learning disabled students; and
- recruitment of mathematics and science teachers and improvement of their skills.

Most of these initiatives are too recent to have been evaluated, and it is difficult to say which are effective and which not. CCSSO is developing a set of indicators for mathematics and science education to allow State-by-State comparison as well as national evaluation of trends.[9]

There is increasing corporate support of K-12 mathematics and science education, but its overall amount remains very small compared with public spending.[10] Industry can also contribute valuable resources in kind, such as equipment, trained scientists and engineers, and site visits. Much of this attention is driven by industrial concerns about the poor quality of high school graduates—the entry level work force to many firms—rather than the question of who will become scientists and engineers.

Course and Curriculum Requirements

States are trying to control and expand what students learn by means of curriculum requirements, tightened graduation requirements, and encouragements to teachers and students to address higher order thinking skills. With respect to graduation requirements, anecdotal data suggest that college admission requirements may be more important than State policies for the college-bound in science and engineering. Indeed, the trend toward tightening graduation requirements may actually be deleterious for this group, by stretching existing teaching resources in mathematics and science too thinly.

Many States have begun to issue reasonably detailed curriculum guidelines, and a few, such as California, have comprehensive guides built around an integrated approach to curriculum development, textbook adoption, and teacher training. Curriculum guides in mathematics and science are used by 47 States; most of these guides are not actually mandatory for school districts. Policies on the amount of time that should be devoted to mathematics and science in elementary schools have been adopted by 26 States. Of these States, most recommend, but do not require, that about 100 to 150 minutes per week be spent on K-3 science, and 225 to 300 minutes per week be spent on K-3 mathematics. For grades four to six, normal recommendations are 175 to 225 minutes per week on science and 250 to 300 minutes per week on mathematics. (See table 6-2.)[11]

All but seven States (and all but four of the fully constituted States) set formal requirements for the award of a high school graduation diploma. (See table 6-3.) (The Constitution of Colorado explicitly forbids the State from setting such requirements.) Almost all of the States that do set formal requirements have, since 1980, steadily increased the number of mathematics and science courses that students must take. Of 47 States that set requirements, 36 require 2 courses in mathematics and 39 require 2 courses in sciences. Delaware, Florida, Guam, Kentucky, Louisiana, Maryland, New Jersey, New Mexico, Pennsylvania, and Texas each set a higher standard, requiring at least one more mathematics course for a total of three.

---

[9]For the pitfalls of making and interpreting State-by-State comparisons of student performance, see Alan L. Ginsburg et al., "Lessons From the Wall Chart," *Educational Evaluation and Policy Analysis*, vol. 10, No. 1, spring 1988, pp. 1-12.

[10]A recent estimate is that total spending in 1986 by 370 companies was about $40 million (6 percent of total corporate spending on education), up from $26 million in 1984. See Council for Aid to Education, *Corporate Support of Education 1986* (New York, NY: February 1988); Anne Lowrey Bailey, "Corporations Starting to Make Grants to Public Schools, Diverting Some Funds Once Earmarked for Colleges," *The Chronicle of Higher Education*, Feb. 10, 1988, pp. A28-30; and Ted Kolderie, "Education That Works: The Right Role for Business," *Harvard Business Review*, September/October 1987, pp. 56-62.

[11]Blank and Espenshade, op. cit., footnote 8, table 1. There is no sound estimate of the average length of the elementary school day available. In 1984-85, it was estimated that the average public school day in elementary and secondary schools included about 300 minutes (5.1 hours) of classes. U.S. Department of Education, Office of Educational Research and Improvement, Center for Education Statistics, *Digest of Education Statistics 1987* (Washington, DC: U.S. Government Printing Office, May 1987), table 89.

Table 6-2.—Comparison of Recommended and Actual Amounts of Time Devoted to Mathematics and Science in Elementary Schools

	Teacher estimates of average number of minutes per day spent on subject		
Grades/subjects	Actual in 1977	Actual in 1986	Recommended
Mathematics:			
K-3	38	38	45-60
4-6	44	49	50-60
Science:			
K-3	19	19	20-30
4-6	35	38	35-45

NOTE: There is no estimate of the average length of the elementary school day available. In 1984-85, it was estimated that the average public school day in elementary and secondary schools included about 300 minutes (5.1 hours) of classes. U.S. Department of Education, Office of Educational Research and Improvement, Center for Education Statistics, *Digest of Education Statistics 1987* (Washington, DC: U.S. Government Printing Office, May 1987), table 89.

SOURCE: Actual amounts of time from Iris R. Weiss, *Report of the 1985-86 National Survey of Science and Mathematics Education* (Research Triangle Park, NC: Research Triangle Institute, November 1987), table 1, p. 12. Recommended times from Rolf Blank and Pamela Espenshade, *State Education Policies Related to Science and Mathematics* (Washington, DC: Council of Chief State School Officers, State Education Assessment Center Science and Mathematics Indicators Project, November 1987), p. 2.

However, control of the *number* of mathematics or science courses alone is a relatively blunt policy tool, for it disregards the curricular *content* of those courses. In addition, mandating extra courses in mathematics and science will be of little use if there are too few well-qualified teachers available to teach them. Indeed, some argue that increasing graduation requirements may actually harm the college-bound in science and engineering, for teachers of the specialized courses that these students now take will be transferred to teach more mainstream courses.[12] Schools with already poorly equipped science laboratory facilities will be asked to spread thin resources even thinner. Thus, tightening graduation requirements must be part of a balanced strategy that also provides adequate teaching and facilities for the new mandated classes in mathematics and science.

Even where States set mandatory minimum requirements, schools and school districts may set

[12]Ben Brodinsky, *Improving Math and Science Education* (Arlington, VA: American Association of School Administrators, 1985), pp. 7-8. Some argue that the trend toward increased control over classroom teaching and learning is both a distinguishing trait of American education and a major weakness. See Arthur E. Wise, "Legislated Learning Revisited," *Phi Delta Kappan*, January 1988, pp. 328-333.

Table 6-3.—Recommended Number of Courses in Mathematics and Science Needed for High School Graduation, by State (for class of 1987 unless specified)

	Courses for regular diploma		Courses for advanced/honors diploma	
	Math	Science	Math	Science
Alabama (1989)	2	2	3	3
Alaska	2	2		
Arizona	2	2		
Arkansas (1988)	5 combined			
California	2	2		
Colorado	Local board			
Connecticut	3	2		
Delaware	2	2		
District of Columbia	2	2		
Florida	3	3	4	4
Georgia (1988)	2	2	3	3
Guam	3	3		
Hawaii	2	2		
Idaho (1988)	2	2		
Illinois (1988)	2	1		
Indiana (1989)	2	2	4	3
Iowa	Local board			
Kansas (1989)	2	2		
Kentucky	3	2	4	3
Louisiana (1988)	3	3		
Maine (1989)	2	2		
Maryland (1989)	3	2	3	3
Massachusetts	Local board			
Michigan	Local board			
Minnesota	0[a]	0[a]		
Mississippi (1989)	2	2		
Missouri	2	2	3	3
Montana	2	1		
Nebraska	Local board			
Nevada	2	1		
New Hampshire	2	2		
New Jersey (1990)	3	2		
New Mexico	3	2		
New York	2	2	2[a]	2[a]
North Carolina	2	2		
North Dakota	2	2		
Ohio	2	1		
Oklahoma	2	2		
Oregon	2	2		
Pennsylvania (1989)	3	3		
Puerto Rico	2	2		
Rhode Island (1989)	2	2	3	2
South Carolina	3	2		
South Dakota (1989)	2	2		
Tennessee	2	2	3	3
Texas	3	2	3	3
Utah (1988)	2	2		
Vermont	5 combined			
Virginia (1988)	5 combined			
Virgin Islands	2	2		
Washington (1989)	2	2		
West Virginia	2	2		
Wisconsin	2	2		
Wyoming	Local board			

[a]New York State Regents courses for credit toward Regents diploma. Minnesota has no State requirements for grades 10-12, 1 math and 1 science required for grades 7-9.

KEY: Combined = 3 mathematics and 2 science or 2 mathematics and 3 science; Local board = requirements determined by local school boards.

SOURCE: Rolf Blank and Pamela Espenshade, *State Education Policies Related to Science and Mathematics* (Washington, DC: Council of Chief State School Officers, State Education Assessment Center Science and Mathematics Indicators Project, November 1987), table 2.

higher standards. Most of those States that do not have minimum requirements set recommended graduation requirements and apply strong pressure on districts to abide by them; some States, such as Michigan, even offer financial incentives to those that do.

Several States, including Indiana, Kentucky, Idaho, Virginia, Texas, and Missouri, now offer advanced or honors diplomas designed explicitly for the college-bound, that require additional coursework or demonstration of competence (again see table 6-3). New York has long offered a "Regents" examination, designed for the college-bound. Data suggest that this examination is effective in encouraging students to take more preparatory mathematics and science courses than is common in other States.[13]

Assessment of What Students Learn

States are making efforts to ensure that teachers address higher order thinking skills in science and mathematics teaching, either through teacher training programs, curriculum frameworks, or through competency testing programs. The Missouri Mastery and Achievement Test, for example, has been designed to include items that assess higher order thinking skills. Higher order thinking, although much sought after, is difficult to define and there appears to be little agreement on how it can be taught.

Statewide testing programs are used in 46 States, indicating a broad response to the public pressure for accountability. But only 30 of these States include science knowledge in these tests, whereas 43 include mathematics. Five States (Alaska, Montana, Nebraska, Ohio, and Vermont) delegate responsibility for assessment to school districts or schools themselves. A recent OTA survey found that 21 States now require students to pass a minimum competency test in designated basic skill areas prior to graduation. Fifteen States include mathematics in such tests and five include science. CCSSO found that 30 States either have, or are planning, competency tests in mathematics, and 6 States in science.[14]

[13]Penny A. Sebring, "Consequences of Differential Amounts of High School Coursework: Will the New Graduation Requirements Help?" *Educational Evaluation and Policy Analysis*, vol. 9, No. 3, fall 1987, pp. 258-273.

[14]U.S. Congress, Office of Technology Assessment, "State Educational Testing Practices," Background Paper, NTIS #PB88-155056, December 1987.

FEDERAL INVOLVEMENT

State and local programs have long been supplemented by Federal efforts.[15] Although these programs have been controversial politically, because of the constitutional limitations on Federal involvement in education, many in mathematics and science education have been reasonably successful in meeting their stated objectives. They are reviewed below.

Given the generally accepted importance of education to the national economy, the Federal Government has long had not only an interest in education issues, but also a mandate to redress inequities in access and provide opportunities for various disadvantaged groups. From the time of the Northwest Ordinance of 1787, the Government has indirectly supported education through financial and land contributions. In 1862, when the U.S. Office of Education was established, that role was augmented by the task of information gathering, research, and analysis. At the turn of the century and for the next 20 years, in response to the increasing industrialization of the Nation, the Federal Government began to take an interest in manpower needs and training and, under Federal law, sought to promote vocational training.

[15]See Deborah A. Verstegen, "Two Hundred Years of Federalism: A Perspective on National Fiscal Policy in Education," *Journal of Education Finance*, vol. 12, spring 1987, pp. 516-548.

The Federal Interest in K-12 Mathematics and Science Education

Large-scale Federal funding of basic research began after World War II, when demand for research scientists and engineers was strong. NSF, created in 1951, was given a mandate to ensure the adequacy of science and engineering education and manpower at all levels. NSF's prime education goal, however, was the cultivation and education of enough scientific talent to fuel the new research-intensive industries and national laboratories. Ever since, NSF has made a significant contribution, by leadership and funding, to science education, but skewed toward students destined for science and engineering careers.

Serious concerns about the adequacy of mathematics and science education developed in the mid-1950s, as Cold War competition with the U.S.S.R. and the baby boom population strained educational resources. Significant NSF involvement in science education, however, originated with the "Sputnik crisis" of the late-1950s. The National Defense Education Act of 1958 was a bold new law to bolster supplies of scientific and other skilled manpower. The act gave grants to school districts to improve or build laboratory facilities and sponsor teacher training in mathematics and science, as well as foreign language instruction. Congress increased funding for science education at NSF, to the point where education was apportioned at about one-half of NSF's budget.[16] The 10 years following Sputnik were the "golden years" of NSF's mathematics and science education effort, highlighted by funding for teacher training institutes, curriculum development, informal education, and research participation programs for high school students and their teachers.

Federal funding for education in all subjects reached its zenith in the early-1960s, when anxiety about regional, economic, and racial disparities in educational provision led Congress to adopt ambitious programs of support for underprivileged students. The Federal role in promoting equity became generally accepted as a consequence of the implementation of these programs, which were successful in achieving their limited goals.[17]

The Reagan Administration's policy of diminishing, if not removing, Federal involvement in education was enacted by major cuts in education programs at the beginnning of the 1980s. Federal support fell from about 8 percent to 4 percent of total national education expenditures. The "new federalism" ideology held that funds were to be apportioned among States on an entitlement basis; States should be allowed to spend funds as they saw fit, and not necessarily in accord with any Federal policies or programs.[18] In 1981-82, the Science and Engineering Education Directorate (SEE) of NSF was disbanded. However, this latter move was most unpopular with mathematics and science educators and in 1983, under pressure, the SEE Directorate was resuscitated.[19]

By 1984, continuing anxiety about international economic competitiveness and the apparently poor quality of public schooling led Congress to pass the Education for Economic Security Act (EESA), designed to promote teaching of mathematics, science, and foreign languages. Title II of this act directs the Department of Education to provide grants to school districts and States to improve the teaching of mathematics, science, computer science, and foreign languages that are critical to national economic well-being. Although the Administration has proposed extending the criteria for funding under this program to all subject areas, this proposal has not been supported in Congress. Title II was part of the package of education programs reauthorized in 1988; the new name for Title II is the Dwight D. Eisenhower Mathematics and Science Education Act.[20]

The Federal Division of Labor in Science Education

Today, Federal mathematics and science education programs are enjoying a resurgence of

---

[16]Myron J. Atkin, "Education at the National Science Foundation: Some Historical Perspectives, An Assessment, and A Proposed Initiative for 1989 and Beyond," testimony before the House Subcommittee on Science, Research, and Technology of the Committee on Science, Space, and Technology, Mar. 22, 1988.

[17]Michael S. Knapp et al., *Cumulative Effects of Federal Education Policies on Schools and Districts: Summary Report of a Congressionally Mandated Study* (Menlo Park, CA: SRI International, January 1983).

[18]See Paul Peterson, *When Federalism Works* (Washington, DC: The Brookings Institution, 1987).

[19]The power of the science education lobby is described in Morris H. Shamos, "A False Alarm in Science Education," *Issues in Science & Technology*, vol. 4, spring 1988, pp. 65-69.

[20]For a synopsis of elementary and secondary education legislation introduced in fiscal year 1988, including that targeted to science and engineering education, see U.S. Congress, Congressional Research Service, *Major Legislation of the Congress* (Washington, DC: U.S. Government Printing Office, August 1988), Issue No. 3, pp. MLC-016 - MLC-021.

funding and considerable bipartisan support, even in the current stringent budgetary climate. They come from three sources:

- NSF, which is generally recognized as the lead agency for mathematics, science, and engineering education;
- the Department of Education, which has overall charge of Federal education programs, but within which mathematics and science education is a relatively low priority; and
- mission agencies that have a direct interest in developing a pool of skilled scientific talent; such agencies include the Department of Energy (DOE), the Department of Agriculture (USDA), the National Institutes of Health (NIH), and the National Aeronautics and Space Administration (NASA).

There is an important difference between NSF and the Department of Education in the scale of the programs that each mounts. Whereas NSF's entire budget is now somewhat under $2 billion annually (and is principally spent on research), that for the Department of Education is about $20 billion. Although most of the Department's programs provide funding either on a categorical basis to providers of education or to ensure equitable access to education, the Department has one program specifically addressed to K-12 mathematics and science education: Title II of the EESA. Appropriations for Title II have been comparable to NSF spending on precollege education, but represent only a few percent of the entire spending of the Department of Education. The dollars, however, are distributed to the States in a formulaic way, with little technical assistance to implement their use or monitor their impacts. No special interest in programmatic mathematics and science education at the Department of Education is apparent.

Some argue that mathematics and science education programs conducted by NSF might flourish were they relocated in the Department of Education. Proponents say that the Department of Education would not concentrate only on the brightest and best students and on funding proposals from rich research universities the way NSF does.[21] In addition, some note that NSF is most comfortable in dealing with and through universities and, until recently, has made few efforts to work closely with States and school districts, as favored by the Department of Education. NSF is now attempting to improve its working relationships with the States. Giving a greater role to the Department of Education in mathematics and science education programs is rejected by most of NSF's existing clients in the scientific and science education research communities, among whom the Department of Education has little credibility.[22]

Consideration of the propriety of Federal education programs is, at root, a highly ideological battle and invokes constitutional concerns. From a public policy perspective, it is important to examine whether and how Federal funding of mathematics and science programs changes the actions that State and local bodies would otherwise have taken. Do Federal programs merely replace funds that would otherwise have been raised by State and local sources or do they allow States and local school districts to do things that they would not or could not otherwise do? Conversely, do Federal programs merely encourage States and local school districts to avoid reforming their own operations, including possibly raising local and State taxes?

There are no clear answers, but Federal programs may work best when they identify aspects of the K-12 education system that have fallen through the cracks of the different agencies involved in education, and address those aspects directly. For example, the "Great Society" legislation of the 1960s improved educational provision to poor and disadvantaged children, and the National Defense Education Act was successful in supplying science equipment and teacher training to school districts.

[21]This is addressed by Don Fuqua in his personal conclusions following hearings by the Science Policy Task Force, *American Science and Science Policy Issues*, Chairman's Report to the Committee on Science and Technology, U.S. House of Representatives, 99th Congress, 2nd Session, December 1986, pp. 80-84 (Committee Print).

[22]See Atkin, op. cit., footnote 16.

NSF's Role in Science and Engineering Education

NSF has been the lead Federal agency in precollege science and mathematics education since the agency's inception. NSF in recent years has been ambivalent towards science education.[23] With its limited funding of K-12 programs, NSF aims to be a catalyst for interplay between the research community and schools and school districts—to generate new ideas for others to implement; to leverage its funding through States, school districts, and foundations; and to do research on mathematics and science education. In its 1987 study, SRI International suggested that NSF's approach to K-12 mathematics and science education should be based on three principles:

- identifying targets of opportunity that are important problems and amenable to NSF's influence;
- supporting core functions of professional exchange among scientists, engineers, teachers, and education researchers; data collection; and experiments in education; and
- investing as part of a coherent strategy to broaden the base of science learners.

The SRI report identified six characteristics of the SEE Directorate that distinguish it from other actors:

- national purview of problems and solutions;
- quasi-independent status rather than an executive branch department;
- connection to the mathematics, science, and engineering communities;
- large amounts of discretionary funding (i.e., those that are allocated on a project/proposal basis);
- a central position vis-a-vis various actors involved in improving mathematics and science education; and
- an established track record in K-12 mathematics and science education programs, especially at the secondary level.[24]

NSF also attempts to define a leadership role in mathematics and science education issues, a function which, especially in the decentralized American education system, should not be underestimated.[25] The ongoing challenge for NSF in science and engineering education will be locating particular niches where it can make a useful and detectable difference, rather than being an agent of change on a broad scale. Investments by the SEE Directorate in science and mathematics education must differ from support for science and engineering research. Developing NSF's capability to invest strategically in education programs may take 5 to 10 years.[26]

Evaluation of Federal Science and Mathematics Education Programs

Too little is known about the effectiveness of previous and current Federal efforts to improve elementary and secondary mathematics and science education. Evaluation of these efforts, for a number of reasons, is very difficult. (See box 6-B.) This review has been supported by an informal questionnaire survey of members of the National Association for Research in Science Teaching (NARST), an association of university researchers in mathematics and science education that aims to improve teaching practices through research. Many Federal programs in this area have been tried, and appendix C lists some of them. Most have emphasized science rather than mathematics.

Three principal Federal programs have been designed to improve mathematics and science education:

- NSF-funded teacher training institutes;
- NSF-funded curriculum development; and
- Title II of the EESA, administered by the Department of Education.

[23] See Deborah Shapley and Rustum Roy, *Lost at the Frontier: U.S. Science and Technology Policy Adrift* (Philadelphia, PA: ISI Press, 1985), pp. 109-114; J. Merton England, *A Patron for Pure Science: The National Science Foundation's Formative Years, 1945-1957* (Washington, DC: U.S. Government Printing Office, National Science Foundation, 1982), ch. 12; and U.S. Congress, Office of Technology Assessment, *Educating Scientists and Engineers: Grade School to Grad School*, OTA-SET-377 (Washington, DC: U.S. Government Printing Office, June 1988), pp. 104-107.

[24] Knapp et al., op. cit., footnote 5, pp. 9-10.

[25] Atkin, op. cit., footnote 16, p. 10.

[26] Knapp et al., op. cit., footnote 5, p. 5.

Box 6-B.—Problems in Evaluating Science Education Programs

Much of the current literature and data on the effectiveness of previous Federal efforts in science education is of poor quality and has limited utility for policy purposes. To be useful for policy, research must measure, quantitively or qualitatively, the effects of a particular program upon a prior situation. It must, therefore, measure what was there before, what happened after, and describe mechanisms by which the addition of a program led to changes apparent by the time of the final observation. Ideally, research should also address the relation between the costs of the program (in money, time, and effort) and its results.

Shortcomings of current research have two origins: 1) the fundamental scientific difficulty of defining and conducting good educational evaluation studies, and 2) idiosyncratic problems with samples and interpretations of research.

The fundamental difficulty is that there is no consensus on what attributes of students should be considered definitive "output" or "input" measures to an educational intervention. For example, many studies measure the student's achievement on multiple-choice tests. Other valid output measures might equally be related to discipline and behavior in school, attitudes and interest in the subject, manual skills, the ability to achieve higher order thinking and reasoning about problems, and the extent to which students feel that they have mastered science and feel confident about it. At the moment, there is broad agreement that standardized achievement test scores should not be used as definitive measures of the outputs of education, but there is no consensus on what combination of other output measures should be used instead.

Such difficulties aside, practical research in the literature often fails to explain fully the effects of programs. Studies of programs that identify highly able children and educate them apart from their peers often show that these children's achievement scores increase more rapidly than those of their age peers. To determine the contribution to achievement scores added by the program, account must be taken of the differential in the rise of scores (due, for example, to maturation) that would have occurred in any case; observations on a control group should be part of the analysis.

In addition, problems in reaching conclusions that are of national relevance arise from the difficulty evaluators have in gaining access to already-beleaguered schools to study programs, and from the expense of doing national evaluations of the worth of particular programs. Schools are increasingly reluctant to allow researchers access to classrooms, since they feel that they are asked to provide too much information already, with too much of it being used against them. These factors sometimes lead researchers to use small samples chosen from unrepresentative school environments for their studies, which diminish the force of research findings. Sometimes researchers argue that the results can be extrapolated to much larger populations, but that argument can only be sustained when other characteristics of the populations, such as socioeconomic status, ethnic composition, and urbanicity of the school chosen, are matched.

NSF Teacher Training Institutes Program

Between 1954 and 1974, NSF spent a total of over $500 million (or over $2 billion in 1987 dollars) on teacher training institutes, most of them for secondary school teachers.[27] These institutes came in several forms. Most were full-time summer programs, others part-time after-school programs, and others full-time academic year programs. Some were aimed at high school teachers, others at elementary and middle school teachers, and others at science supervisors. At that time, the consensus was that the teaching force was deficient in content knowledge about science, and most of the institutes focused on improving teachers' knowledge about science, largely by means of lectures. Few of them addressed teaching practices. Most institutes were based at colleges and universities, which organized the programs and paid teachers stipends and travel expenses for attending.

By today's standards, the model that these institutes adopted—that giving mathematics and science teachers better subject knowledge would lead to better teaching—seems rather primitive. Any replication of these institutes would now focus on achieving a balance between knowledge and experience of *how* to teach science and mathematics together with *what* to teach. Nevertheless, evi-

[27]Ibid., vol. 1, p. 133. Expression in 1987 dollars is an OTA estimate.

dence suggests that the former mathematics and science teaching force was so poorly endowed with subject knowledge that emphasis on content over practice was largely inevitable.

At their peak in the early-1960s, about 1,000 institutes were offered annually, each with between 10 to 150 teachers meeting over a 4- to 12-week period; just over 40,000 teachers were reached annually (about 15 percent of the high school mathematics and science teaching force). The institutes were reorganized in 1970, by which time NSF had judged that the institutes had reached about as many teachers as any voluntary program ever would. A survey taken in 1977 found that many teachers had participated in NSF-funded institutes. The survey indicated that nearly 80 percent of mathematics and science supervisors had attended an institute, as had 47 percent of science teachers and 37 percent of mathematics teachers of grades 10 to 12. Only about 5 percent of teachers of kindergarten through third grade had attended such an institute.

Improvement of the skills of the elementary teaching force will always be difficult, because there are over 1 million individuals who teach at least some elementary mathematics and science (along with many other subjects), and because university-based mathematics and science educators generally find it harder to reach elementary teachers than they do high school teachers. For example, it is estimated that NSF's current efforts reach, at most, 2 or 3 percent of secondary mathematics and science teachers.[28]

Evaluation of the effects of these institutes has yielded no consensus on their usefulness. Teachers who participated are enthusiastic about them and remember them as stimulating and professionally refreshing. The General Accounting Office reviewed research on NSF-funded institutes and found little or no evidence that such institutes had improved student achievement scores.[29] Whereas NSF remains cautious in claiming effectiveness, many science educators think that the institutes were very successful. Various studies, including one by the Congressional Research Service in 1975, found that the institutes had positive effects.[30]

NSF did not attempt a systematic comprehensive evaluation of its own during the lifetime of the institutes. The institutes concentrated on improving the mathematical and scientific knowledge of teachers, but there is no direct relationship between teachers' knowledge, their effectiveness as teachers, and educational outcome measures such as students' achievement test scores.[31] In practice, teachers need both some knowledge of mathematics and science and some pedagogical skills to be effective teachers.

Anecdotal evidence drawn from a history of this program and from the OTA survey of NARST members indicated that the teacher institutes program had these important effects:

- It brought teachers up-to-date with current developments in science.
- It brought teachers closer to the actual process of doing science and thereby improved both their identification with, and sense of competence in, science.
- It helped teachers share common solutions and problems, and gave them a network of peers that they kept in touch with many years after the programs ended.
- It allowed teachers to do experimental work in science, which many of them had never done before, and thereby encouraged them to replicate this experience for their students.
- It helped define leaders for the science education community, who now are effective voices for this community in professional meetings and policy debates.

---

[28]Ibid., vol. 1, pp. 133-134.

[29]Such a *post hoc* measure, however, may bear no relation to the *a priori* goal of the institutes, namely, to update teacher's knowledge. See U.S. Congress, General Accounting Office, *New Directions for Federal Programs To Aid Mathematics and Science Teaching*, GAO/PEMD-84-5 (Washington, DC: U.S. Government Printing Office, Mar. 6, 1984).

[30]U.S. Congress, Congressional Research Service, *The National Science Foundation and Pre-College Science Education: 1950-1975*, Committee Print of the U.S. House of Representatives Committee on Science and Technology, Subcommittee on Science, Research, and Technology, January 1976; Hillier Krieghbaum and Hugh Rawsom, *An Investment in Knowledge* (New York, NY: New York University Press, 1969); and Victor L. Willson and Antoine M. Garibaldi, "The Association Between Teacher Participation in NSF Institutes and Student Achievement," *Journal of Research in Science Teaching*, vol. 13, No. 5, 1976, pp. 431-439.

[31]Knapp et al., op. cit., footnote 5, vol. 1, p. 130.

- It recognized the importance of the work of mathematics and science teachers.
- It inspired and invigorated teachers to face another year of teaching.

Among problems with the institutes, however, were the following:

- Since teacher participation was voluntary, only those teachers who were the most interested and motivated in science teaching participated. The least interested and least qualified shunned the program, which consequently reached at best only about one-half of all teachers (though this is not an insignificant number).
- Since many of the institutes offered graduate credit, the program helped subsidize students' progress toward master's and doctoral degrees in education. Possession of these degrees may have helped impel good teachers out of active teaching into administration, or into other jobs entirely.
- The institutes succeeded in conveying much information about new developments in science, but gave teachers few clues as to how to teach this information to their classes.
- The institutes were typically lecture courses.

The costs of the teacher institutes were large, particularly for the academic year institutes, for which teachers were pulled from classrooms. Typical costs today for teacher institute programs are reported at about $25 to $40 per hour per teacher. If a minimum of 100 hours is assumed for length of the institute to make it meaningful, for a total cost per teacher of about $3,000, it would cost about $600 million to put all secondary mathematics and science teachers through one institute each.[32]

A small program of teacher institutes is now funded through the Teacher Preparation and Enhancement Program at NSF. In general, these put more emphasis on teaching techniques in mathematics and science than did the earlier institutes. Evaluations of these institutes indicate that they are having some success.[33]

Any future replications could build on successful NSF-sponsored models for inservice education, and would need to be based on a stronger partnership between school districts and universities than existed 20 years ago. Alternatively, NSF could try to act as a catalyst, persuading States and school districts to fund such inservice education directly.[34] Either way, the *school* must be recognized as the appropriate unit; teachers alone are neither the problem nor the solution. Organizational change is needed, and that impinges on all aspects of the system, from the classroom and school administration to the local district and State jurisdictions.

NSF-Funded Curriculum Improvements

In the 1960s and 1970s, NSF spent about $200 million on over 50 curriculum reform efforts. Their main focus was on the sciences rather than mathematics, and the new curricula have had a significant impact on the science education of many students.

To protect against Federal domination of curriculum, NSF did not review these projects once completed and ensured that the materials were held and disseminated by the developer rather than by NSF. Nevertheless, a fifth grade social studies curriculum, "Man: A Course of Study," attracted considerable criticism and congressional scrutiny in 1975; some critics found it offensive and unacceptable for children. NSF has since been extremely cautious in curriculum development ever since.

The curricula were almost all developed by teams headed by scientists, but often involved mathematics and science educators, and general educators; they were largely designed to convey the content and structure of the separate scien-

[32]Ibid., vol. 1, p. 132.

[33]Renate C. Lippert et al., "An Evaluation of Classroom Teaching Practices One Year After a Workshop for High School Physics Teachers," unpublished paper, May 1987; and Margaret L. While et al., "Biosocial Goals and Human Genetics: An Impact Study of NSF Workshops," *Science Education*, vol. 71, No. 2, 1987, pp. 137-144.

[34]Or the National Science Foundation could concentrate on developing a smaller number of lead mathematics and science teachers who would then enthuse the remaining teacher force. This latter strategy is advocated in the recent SRI report, primarily because of the cost and difficulty of organizing another mass program to reach all mathematics and science teachers. Such a program, however, would focus change one further step away from students' actual learning about mathematics and science than even the teacher institute program would. See Knapp et al., op. cit., footnote 5, p. 135.

tific disciplines.[35] Many believed that the high school curricula were too demanding, however, and worked well only for those planning college majors in science and engineering. Many believed that the scientists dominated the projects and that feedback on the new materials was not sought.

In 1968-69, it was estimated that nearly 4 million students (about 10 percent of the total) were using some kind of NSF-funded curriculum materials. A 1977 survey found that about one-half of the science classes in grades 6 to 12 were using such materials, although only about 10 percent of mathematics classes in these grades were.[36] NSF recently reinstated curriculum development, and is funding a series of three-way collaborative projects (involving publishers, universities, and school districts) to develop elementary science curricula.

Research reflects a consensus that the new curricula worked quite well. A review of evaluation studies found that, compared to control groups, students taking new curricula scored higher on achievement tests, had more positive attitudes toward science, and exhibited fewer sex differences in these attributes. Students taught by teachers who had been through preparatory teacher institutes for the curricula scored more highly than their peers taking the new curricula without this benefit.[37] It is clear that successful implementation of new curricula is very closely tied to teacher training; curriculum projects are of little use without support for teachers to master the new curriculum and put it into practice. Earlier elementary science education curricula, such as the Science Curriculum Improvement Study (SCIS), Elementary Science Study (ESS), and Science: A Process Approach (SAPA) did not become well established in schools largely because teachers were ill-prepared to teach them.[38]

Among the positive effects of new curricula were a spill-over effect to the entire mathematics and science curriculum, such that traditional textbooks from commercial publishers began to adopt many of the techniques, such as hands-on science. The main NSF-funded mathematics curriculum, the School Mathematics Study Group, explicitly aimed at affecting publishers' own curricula.[39]

The success of new curricula is partially attributed to the considerable research that went into them. Many are still in use in some schools; they have influenced teachers and textbooks ever since. Many teachers have extensive experience with these curricula, and now know how to avoid some of the problems that the curricula can cause.

NARST members particularly cited the Biological Sciences Curriculum Study (BSCS), the ESS, the SAPA, and the SCIS (the last three being elementary science curricula), as effective and valuable curricula. Many of these curricula have apparently been translated and adopted for use in schools in South Korea and Japan. It has been estimated that the BSCS has been used in about one-half of all biology classes in the United States. One-quarter of those who graduated with a baccalaureate degree in physics in 1983-84 had taken BSCS physics in high school.[40]

Problems cited with these curricula included:

- A lack of adequate financial and moral support to teachers introducing the new curricula.
- A focus on "pure" science rather than its real-life applicability.
- A design with only the future scientist and engineer in mind, which frustrated large numbers of mainstream students.
- A domination by research scientists rather than science educators, precluding develop-

[35]Ibid., vol. 1, p. 92.

[36]Iris R. Weiss, *Report of the 1977 National Survey of Science, Mathematics, and Social Studies Education* (Research Triangle Park, NC: Center for Educational Research and Evaluation, 1978), p. 83.

[37]A comprehensive analysis of 81 other studies is reported in James A. Shymansky et al., "A Reassessment of the Effects of 60's Science Curricula on Student Performance: Final Report," mimeo, n.d. (a reworking of material originally published in 1983). Also see Patricia E. Blosser, "What Research Says: Research Related to Instructional Materials for Science," *School Science and Mathematics*, vol. 86, No. 6, October 1986, pp. 513-517; and Ted Bredderman, "Effects of Activity-Based Elementary Science on Student Outcomes: A Quantitative Analysis," *Review of Educational Research*, vol. 53, No. 4, winter 1983, pp. 499-518.

[38]Knapp et al., op. cit., footnote 5, vol. 1, p. 68. Ironically, anecdotal evidence suggests that these programs were especially successful with minority and disadvantaged students (Shirley Malcom, American Association for the Advancement of Science, personal communication, August 1988).

[39]Knapp et al., op. cit., footnote 5, p. 48.

[40]Ibid., vol. 1, p. 92.

ment of a needed team approach to the design and implementation of the curricula.

- An approach embracing a view of science as a collection of neutral and immutable facts, to the exclusion of other conceptions of science.

Today's science curricula have slipped back into stressing facts at the expense of reasoning and understanding. One approach to future science curricula might be to disregard traditional disciplinary boundaries in science, and focus on hybrid disciplines or unifying themes and ways of observing and measuring. Many science-intensive schools integrate science with mathematics courses, an innovation that could be followed by most schools. In any case, the design of new curricula requires the active participation of the relevant scientific research communities.[41]

Mathematics curricula also need to be improved. Less emphasis needs to be placed on traditional rote learning. The curricula could concentrate on new developments in mathematics and its applications, such as mathematical problem solving, probability and statistics, and computer science. Deficiencies of mathematics curricula have been demonstrated by data collected for the international comparisons of achievement.[42] Curricula need to reflect the availability of the hand-held calculator. One example of a reform in progress is the new Mathematics Framework for California Public Schools, which encourages calculator use from primary grades upward.

Title II of the Education for Economic Security Act of 1984

Title II of the EESA (Public Law 98-377 as amended by Public Law 99-159 and Public Law 100-297) was a major congressional initiative to address the problems apparent in mathematics and science education in the early-1980s. It provides funds to both States and school districts to improve the skills of teachers and the quality of instruction in mathematics, science, computer learning, and foreign languages in both public and private schools. Title II established teacher training as first priority and directed that the funds allocated to school districts must be spent on training. Only if school districts demonstrate to States that there is no further need for such training can such funds be used for other purposes, such as equipment and materials purchases or training in foreign language or computer instruction.[43] The legislation also contains provisions intended to boost the participation of "underrepresented" and "underserved" groups. The legislation was reauthorized in 1988 (Public Law 100-297), and renamed the Dwight D. Eisenhower Mathematics and Science Education Act.

Title II has been funded unevenly by Congress: $100 million was appropriated in fiscal year 1985, $42 million in fiscal year 1986, $80 million in fiscal 1987, and $120 million in fiscal year 1988. For comparison, the total expenditure on public and private elementary and secondary education in 1986 was about $140 billion; the total spending on mathematics and science teacher training was probably about $500 million to $1 billion. Total spending by the Department of Education was $19.5 billion in fiscal year 1987 (so that Title II was a small portion of the Department's total effort). Another way of evaluating spending on Title II is to recall that, nationally, a $40 million education program equates to a spending of $1 per pupil *or* $20 per teacher.

The Department of Education's implementation of Title II has been slow. Although funds for fiscal 1985 were provided by Congress, grant awards were not announced until July 2, 1985, immediately after the end of the school year (effectively delaying implementation by 12 months). However, the Department now hosts regular meetings of State Title II coordinators and publishes some information on exemplary State and local pro-

[41]See Philip W. Jackson, "The Reform of Science Education: A Cautionary Tale," *Daedalus*, vol. 112, No. 2, spring 1983, pp. 143-166.

[42]Curtis C. McKnight et al., *The Underachieving Curriculum* (Champaign, IL: Stipes Publishing Co., January 1987); and Knapp et al., op. cit., footnote 5, vol. 2, pp. 35-60.

[43]Even then, no more than 30 percent of the funds provided to each school district can be used to purchase computers or software, and no more than 15 percent can be used for foreign language instruction. Note also that some of the funds appropriated under the act are spent on these areas via the Secretary's Discretionary Fund (see below).

grams that have been funded through the Title II program.[44]

The legislation specifies in some detail the intended fate of the sums appropriated, leaving relatively little discretion to the Department of Education, the States, or school districts. It is worth examining the provision of the legislation in some detail to understand the way in which mathematics and science education programs administered through the Department of Education operate. (See figure 6-1.)[45]

Of the funds appropriated under Title II, 90 percent is sent straight to the States as categorical grants. Nine percent is retained by the Department of Education to be spent on "National Priority Programs" in science, mathematics, computer, and foreign language education via the Secretary of Education's Discretionary Fund, and

[44]Carolyn S. Lee (ed.), "Exemplary Program Presentations, Title II of the Education for Economic Security Act," mimeo prepared by the U.S. Department of Education for the December 1987 meeting of Title II coordinators, Washington, DC; and Carolyn S. Lee (ed.), "Exemplary Projects: Mathematics, Science, Computer Learning, and Foreign Languages: A Collection of Projects Funded Through Title II of the Education for Economic Security Act," mimeo prepared for December 1987 meeting of Title II Coordinators, Washington, DC.

[45]The following is based on the allocations that applied during fiscal years 1985 to 1988. They have been amended somewhat in the recent reauthorization of the legislation.

Figure 6-1.—Distribution of Federal Funds Appropriated Under Title II of the Education for Economic Security Act of 1984

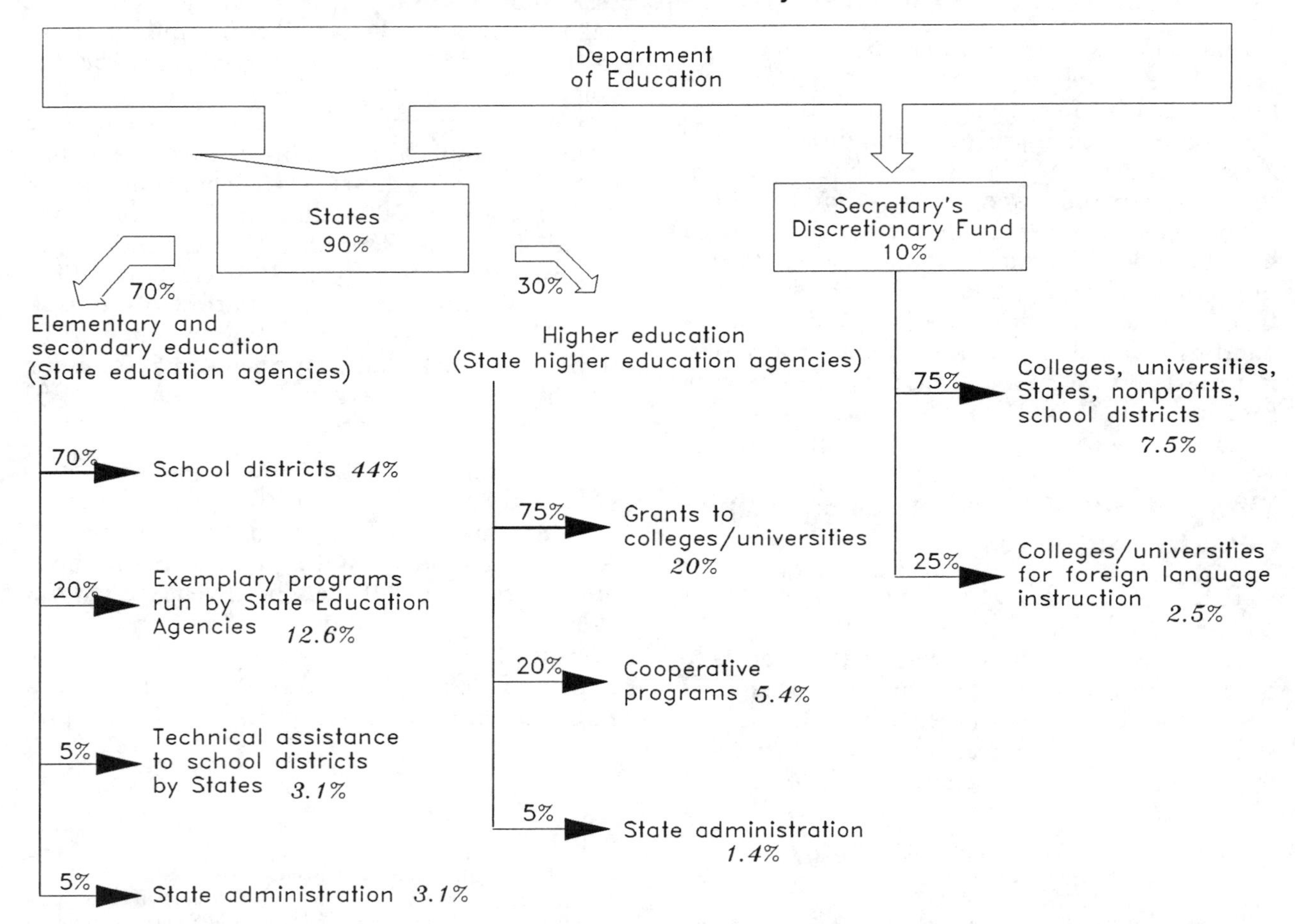

NOTE: Numbers in italics are final percentages of the original Department of Education 100% allocation; the numbers not italicized are distribution formulas.

SOURCE: Office of Technology Assessment, 1988; based on data from the U.S. Department of Education.

the remaining 1 percent is divided equally between the U.S. Territories and Insular Areas and the U.S. Bureau of Indian Affairs. Each State must submit a plan for the use of the funds to the Secretary of Education before the State's allocation can be released.

The States' 90 percent allocation is divided among the States (including Puerto Rico and the District of Columbia) on the basis of the size of their school-age populations, with the proviso that each State must receive at least 0.5 percent of the total appropriation. Of the sum that each State receives, 30 percent must be given directly to the State's higher education agency (primarily for elementary and secondary teacher training programs), leaving 70 percent to be allocated by the State education agency for elementary and secondary education (no more than 5 percent of each of these allocations can be used for State administrative costs). Of the funds given to each State's education agency, 70 percent must be divided among the school districts, giving equal weight to the size of the total public and private school-age population and the number defined as "disadvantaged" under the Chapter 1 program of the Education Consolidation and Improvement Act of 1981. The 30 percent share of funds intended for elementary and secondary education that is retained by the State must be spent on exemplary programs in teacher training and instructional equipment, including those designed specifically to benefit the disadvantaged or the gifted, and on technical assistance to school districts.

Of the sum allocated to the Secretary's Discretionary Fund, 25 percent must be awarded to higher education for use in foreign language instruction improvement, and 75 percent must be spent on competitive awards for special programs. So far, two competitions for this fund have been held, awarding $5 million. In these competitions, special consideration must be given to magnet school programs for gifted and talented students and for historically underserved groups in science and engineering.

The net effect of the Title II legislation is to disperse appropriated funds very widely without consideration of whether dilution reaches a threshold where the funds make no discernible impact on mathematics and science education. Almost all school districts in the Nation have received small amounts of these funds; the problem is the size of the allocation. One-half of all the annual grants made to school districts under Title II were for less than $1,000 and one-quarter were for under $250; some districts have refused to apply for the funds, citing the desultory amounts of money that they would get as a result.[46] Most State Title II coordinators do not collect detailed data on the uses of these funds, but they believe that most are used for inservice training and workshops. Very little goes to support alternative certification and programs for training new teachers or teachers switching into science and/or mathematics.[47]

The legislation required States to justify their needs for this funding by providing needs assessments to the Department of Education, describing their plans for upgrading teacher quality in mathematics, science, computer learning, and foreign languages. Data yielded by these assessments have been of variable quality. CCSSO, with funding from NSF, did attempt to encourage States to report their data using common formulae and tabulations, and about 33 States (mainly the smaller States) supported this program. Congress required the Department of Education to provide it with a summary of the needs assessments in the fiscal 1986 NSF authorization (Public Law 99-159); the Department fulfilled this requirement in September 1987, although this document itself noted that it was of limited usefulness since the ". . . resulting needs assessment reports . . . are highly idiosyncratic and do not readily lend themselves to national generalizations."[48]

[46]The evaluation of the distribution of funds under the program notes that one (unnamed) school district, which would receive $25 under the program, was advised by its State Title II coordinator to discuss the district's inservice training needs for teachers over two and a half cases of beer.

[47]Ellen L. Marks, *Title II of the Education for Economic Security Act: An Analysis of First-Year Operations* (Washington, DC: Policy Studies Associates, October 1986).

[48]Royce Dickens et al., "State Needs Assessments, Title II EESA: A Summary Report," prepared for U.S. Department of Education, Office of Planning, Budget, and Evaluation, August 1987.

Data and Research Funded by the Department of Education and the National Science Foundation

The Federal Government, through NSF and the Department of Education, supports a great deal of data collection and analysis, as well as educational research and evaluation that is relevant to mathematics and science education. The following programs are active:

- Office of Studies and Program Assessment, NSF, providing data and management information on the national state of mathematics and science education;
- Program of Research in Teaching and Learning, NSF, funding educational research on effective teaching and learning in schools and universities;
- Center for Education Statistics, Office of Educational Research and Improvement (OERI), Department of Education, providing national data on education in all subjects; and
- research programs in OERI, funding educational research by individual investigators and centers, and its dissemination through research and development (R&D) centers and databases, such as ERIC.

The overall system of data collection on mathematics and science education would benefit from greater formal coordination between NSF and the Department of Education.[49] The best work has been done by NSF, particularly its two National Surveys of Mathematics and Science Education (in 1977 and 1985-86, respectively). But data on class enrollments by sex, race, and ethnicity are not available from the 1985-86 survey, and preliminary data on course-taking by high school graduates from the National Assessment of Educational Progress 1987 High School Transcript Study appeared late in 1987. A new Department of Education study, the National Educational Longitudinal Study, officially began in 1988. It would be valuable to have recent data to gauge the effects of educational reforms.[50]

Educational research in mathematics and science is in even more tenuous shape, however, having suffered (along with research in other subject areas) from budget cuts,[51] disputes among several relevant disciplines (including psychology and cognitive science), and a failure to pursue more active development programs as well as basic research.[52] In particular, education research in mathematics and science is still recovering from the shutdown of NSF's SEE Directorate in the early 1980s.[53] The key to better research, ultimately, will be better dissemination and development work that builds on the base of empirical knowledge. Data collection from States and school districts should be improved by changes enacted by Congress during the recent reauthorization of Federal education legislation.

The Department of Education also runs the National Diffusion Network (NDN), established in 1973, which disseminates and provides some funding for implementation of curricula that are of demonstrated educational benefit. NDN programs cover all levels of education, including higher education. By December 1987, there were 450 programs in NDN, and they were being used in about 20,000 schools with 2 million students. Ten of these programs were in science, four of which had originally been developed, in part or whole, with funds from NSF. Recent programs had been developed with Title II funds. NDN funds dissemination of programs, and the average grant is about $50,000 over 4 years.[54]

[49]Informal coordination has been practiced for a long time, including joint National Science Foundation (Science and Engineering Education Directorate)-Department of Education review of educational research proposals (Richard Berry, former National Science Foundation staff, personal communication, August 1988).

[50]The Council of Chief State School Officers also coordinates various data collections by the States. Other Department of Education-funded longitudinal studies, the National Longitudinal Study and the High School and Beyond survey, have proved to be invaluable as well.

[51]U.S. Congress, General Accounting Office, *Education Information: Changes in Funds and Priorities Have Affected Production and Quality*, GAO/PEMD-88-4 (Washington, DC: U.S. Government Printing Office, November 1987).

[52]On research needs in science education generally, see Marcia C. Linn, "Establishing a Research Base for Science Education: Challenges, Trends, and Recommendations," *Journal of Research in Science Teaching*, vol. 24, No. 3, 1987, pp. 191-216.

[53]See Knapp et al., op. cit., footnote 5, vol. 2, pp. 1-27 to 1-50. For example, see National Science Foundation, *Summary of Active Awards, Studies and Analyses Program* (Washington, DC: March 1988).

[54]All data are from U.S. Department of Education, Office of Educational Research and Improvement, *Science Education Pro-*

Education Programs in the Mission Research Agencies

Federal R&D mission agencies are active in mathematics and science education. The bulk of activity is in informal outreach programs, such as classroom visits, NASA's Spacemobile, laboratory open houses, career days, and science fairs. These programs touch tens of thousands of students—but only to a small degree. Some Federal laboratories "adopt" a high school, and thereby develop extensive contact with teachers and students. Adoption programs usually involve little or no direct cost to the agency, since they rely on employee volunteers to speak at schools, judge fairs, or host a visiting school group. Because these kinds of programs are informal in nature, they do not have established budgets or staff at the agency, and depend on the initiative of research staff as well as any education coordinators that the agency may have. They tend to wax and wane. Usually only a handful of people at any agency (including its field laboratories) work full time on education.

A few agencies also have modest programs to train and support mathematics and science teachers, ranging from summer research and refresher courses to resource centers where they can work with and copy instructional materials. None of the extensive teacher training workshops reach more than a few dozen teachers, however. One exception is NASA's interactive videoconferences on NASA's activities in space science education, which have drawn about 20,000 teachers each time.

The other genre of program is in research participation or apprenticeships, which reaches only a very few students but in much greater depth. These programs bring students, usually juniors or seniors in high school, into agency laboratories for research experience in ongoing mission R&D.

These research apprenticeships are powerful mechanisms. For example, since 1980, the NIH Minority High School Student Research Apprenticeship Program has awarded grants to universities and medical schools to bring students in for summer research projects. Students participate in all aspects of research; they collect data, review literature, attend seminars, use computers, and write up findings. In 1988, over 400 students will be supported on NIH grants totaling $1.5 million. At the program's peak in 1985-86, 1,000 students participated each summer. Over half the participants are Black; about 20 percent are Hispanic, another 20 percent Asian. Over half are female. Students are selected for aptitude and motivation, and on the recommendation of their science teacher.[55]

Mission agency summer research programs together support perhaps a few thousand high school students—certainly under 10,000—but, in any case, more than NSF's comparable program does. Full-fledged programs cost on the order of just under $1,000 to around $3,000 per student. Most of the cost is in salary for the student; there usually is significant cost-sharing with the host laboratory, and often with industry and other private supporters. Shorter summer programs, where students come in for a few weeks, or programs that mix hands-on research with instruction or career sessions, are less costly per student (up to a few hundred dollars); they also provide a less intensive experience.

Mission agency education activities, mostly local and informal, complement to a large extent the formal, research-oriented and nationwide programs of NSF. The mission agencies are particularly successful at reaching diverse populations. The programs build on the existing agency staff and facilities; many make special efforts to reach females and minorities. Formal minority research apprenticeships were established in 1979 by the Office of Science and Technology Policy for the major R&D agencies.

The mission agencies have unique strengths in reaching out to young science and mathematics

grams That Work: A Collection of Proven Exemplary Educational Programs and Practices in the National Diffusion Network, PIP 88-849 (Washington, DC: U.S. Government Printing Office, 1988).

[55]The National Institutes of Health evaluation testifies to the success of the research apprenticeships in encouraging participants to attend college and pursue a research career; 60 percent of students say that the experience influenced their career decisions. One of the touted strengths of the program is its flexibility and the institution's leeway in awarding and using the grants (the grants include salary for the apprentice, which may also be used for supplies or any relevant education activity). National Institutes of Health, personal communication, May 1988.

students. Federal research laboratories around the country are a particularly valuable resource for education, with unique facilities and equipment. DOE and NASA have large, visible laboratories doing state-of-the-art research in exciting areas such as space flight, lasers, the environment, and energy. Many of these laboratories are in areas where there is little or nothing else in the way of sophisticated research facilities. There is no place else students (or teachers) can see, touch, and work with equipment like rocket engines, whether it is for an afternoon visit or a summer of research. The thousands of researchers at these laboratories are likewise a valuable resource, offering inspiration, expertise on careers and nearly any special area of research, real-life role models for youngsters, and mentors for students doing research.

The ethos behind mission agency education efforts is to improve the quality and coverage of elementary and secondary education in their mission area, and ultimately to help ensure an adequate supply of scientists and engineers working both for the agency and in areas that support the agency's mission. (See table 6-4.)

Major Mission Agency Programs

NASA is one of the most active and innovative mission agencies in education. Public education has been an integral part of NASA since its inception, and is a natural response to the continuing interest of children—and adults—in space science and exploration. The excitement of NASA's space mission clearly is an interesting way to package basic science and mathematics les-

Table 6-4.—Summary of Mission Agency K-12 Mathematics and Science Education Programs

Agency programs	FY 1988 budget	Number of students	Females/ minorities?
Summer research and enrichment:			
DOE High School Honors Research Program	$550,000	320	N
DOE Minority Student Research Apprenticeships	$120,000	200	Y
NIH Minority High School Student Research Apprenticeships	$1.5 M	410	Y
DoD Research in Engineering Apprenticeship			Y
DoD High School Apprenticeship (Office of Naval Research, Army Research Office, Air Force Office of Scientific Research)			Y
USDA Research Apprenticeship (Agricultural Research Service)	$250,000	200	
NASA Summer High School Research Apprenticeship Program		125	Y
General enrichment: [a]			
DOE Prefreshman Engineering Program	$300,000	2,000	Y
DoD/Army Computer-Related Science and Engineering Studies (4 weeks)	$ 50,000	60	
DoD UNITE		500-2,000	Y
NOAA D.C. Career Orientation	$ 30,000	24	Y
EPA summer internships			
USDA 4-H	$70-100 M	5 M	N
Teacher training and support:			
NASA education workshops	$1 M+	unknown	
NASA resource centers		unknown	
DOE research experience and institutes	$250,000	50	
USGS summer jobs for teachers	unknown	20-90	
Informal outreach: [b]			
All agencies (especially NASA and DoD); (also NIST, DOE, USDA, NOAA, USGS, EPA, NIH)		hundreds of thousands	

[a]Research, instruction, career orientation' usually short-term residential.
[b]Classroom visits and demonstrations, science fairs, talent searches, career fairs, laboratory open houses, weekend instruction and hands-on programs, partnerships ("adopt a school"), materials development and lending.
NOTE: Most programs include cost-sharing with host institution, and often with local industry and other sponsors.
KEY: DOE = Department of Energy
NIH = National Institutes of Health
USDA = U.S. Department of Agriculture
NASA = National Aeronautics and Space Administration
DoD = Department of Defense
NOAA = National Oceanic and Atmospheric Administration
EPA = Environmenal Protection Agency
USGS = U.S. Geological Survey
NIST = National Institute of Standards and Technology

SOURCE: Office of Technology Assessment, 1988.

sons. An educational affairs division was formally created in 1986. NASA supports elementary and secondary teachers with workshops and resource materials at NASA's field centers. NASA is also reviewing and supplementing existing curricula, and has produced videotapes, satellite broadcasts, videodiscs, electronic tutors, and other innovative educational technologies. NASA has a summer high school apprenticeship, which employs about 125 students, most of them minority. In addition, about 30 full-time education outreach staff (mostly former teachers) spend over 160 days on the road with the Spacemobile (a mobile resource center for children about the space program), presenting school assemblies and class rooms.[56]

DOE, in cooperation with its national laboratories, has one of the most extensive programs for student summer research among the Federal agencies. It has several programs, for example, the high school honors research program (320 students, $550,000), and minority student research apprenticeships (200 students, $120,000). DOE brings in teachers, offering them research experience and training (in 1988, 50 teachers, $250,000). It is also building science education centers in its national laboratories as part of a university-laboratory cooperative program.

[56]See National Aeronautics and Space Administration, "Educational Affairs Plan: A Five-Year Strategy, FY 1988-1992," unpublished manuscript, October 1987.

Photo credit: Children's Television Workshop

There are many Federal programs which support innovation and research in science and mathematics education.

DOE and the Department of Defense (DoD) are the only agencies with substantial involvement in early engineering education. The prefreshman engineering program (PREP) awards money to universities to sponsor summer programs for females and minorities in junior high and high school. PREP programs include research experience, instruction, and career and college preparation. It reaches 2,000 students ($300,000), and benefits from substantial cost sharing and in-kind support.

Various offices of DoD support UNITE (Uninitiates Introduction to Engineering), which range from short stays to many-week research apprenticeships on engineering campuses. UNITE sessions include engineering and other technical classes, planning for college, career seminars, visits to military laboratories and facilities, and meetings with military engineers. Extensive followup of students shows that the program works. Students report that it helped shape their career goals; it turned some off, but it turned many more on to engineering. (UNITE is a military offshoot of programs sponsored by the Junior Engineering Technical Society, or JETS, a private organization with extensive corporate support. JETS coordinates Minority Introduction to Engineering programs to introduce students to engineering and college, most often as 1- or 2-week summer programs at universities.) The Research in Engineering Apprenticeship Program pays several hundred high school students to do mentored summer research at defense laboratories.

The National Institute of Standards and Technology (NIST), formerly the National Bureau of Standards, is a relatively small agency with few regional facilities. However, it does extensive informal outreach from its major laboratories in Maryland and Colorado. NIST encourages its research staff to give talks and demonstrations at schools, and work with science fairs and informal education programs.

USDA is surprisingly active in early science education. In particular, the well-established 4-H program, which has over 5 million participants in counties throughout the Nation, reaches an enormous number of children. USDA is launching an initiative to bring more basic science content into 4-H programs, involving such modern areas relevant to agricultural research as biotechnology and remote sensing. (See box 6-C.) They have received a small amount of funding from NSF to develop innovative science programs within 4-H. USDA also has many informal outreach and career programs. The Agricultural Research Service is the home for USDA's summer research apprentice program, which supports about 200 students each summer.

Within the Department of the Interior, the U.S. Geological Survey has extensive and well-organized elementary and secondary education programs. They also sponsor summer jobs for geology teachers. Other branches of the Department of the Interior have educational programs, mostly informal outreach. Much of the work of the Interior lends itself to summer internships for students.

Box 6-C.—4-H: Five Million Children

There are nearly 5 million children involved in 4-H, most under the age of 12.[1] The goals of 4-H are to teach children about agriculture and related sciences, increase technological and science literacy, and to interest children in agriculture and science careers.[2] 4-H is mounting a science and technology initiative that will broaden its scope to include the modern technologies and basic biological, physical, and chemical sciences that feed into agriculture. The core of 4-H is hands-on individual projects that allow children to learn by doing. Real-world applications include agriculture, food, nutrition, and soil science, from the more familiar growing gardens and raising goats to computer programming, space-based remote sensing, and molecular genetics. 4-H participants complete several projects a year that teach principles of the scientific method. Other media include instructional TV, science and agriculture fairs, visits to colleges and universities, and camps.

4-H draws upon a vast network of county extension agents, professionals, educators, and volunteers. This network is based in land-grant colleges and extension services. They support the over 600,000 volunteer leader-teachers who assist children in their projects. Many programs are in-school, involving teachers and clubs (4-H provides informal teacher training). Others are out of school with family, employer, and community support.

The national 4-H office is also sponsoring research into how children prepare for and choose agriculture and other science careers. This aims to promote positive attitudes toward science and technology, and helps 4-H design programs to encourage children to enroll in mathematics and science courses in high school, and later to enter science majors at universities (particularly, but not exclusively, agriculture-related majors at land-grant universities).

Money comes from the Cooperative Extension Service, jointly funded by Federal, State, and county governments. Total public funding for 4-H is about $260 million. The Federal contribution is probably about $70 to $100 million, although it is difficult to estimate because all Federal funds go into the general Cooperative Extention Service budget and cannot be distinguished by their end use. Industry and foundation contributions are on the order of $50 million. The estimated value of volunteer time is $1 billion.

[1]There are slightly more girls than boys, and just over 20 percent minority (16 percent Black, 1 percent American Indian, 4 percent Hispanic, and 2 percent Asian and Pacific Islanders).

[2]Allan Smith, U.S. Department of Agriculture, personal communication, February 1988; U.S. Department of Agriculture, Cooperative Extension Service, *Science and Technology: The 4-H Way, Status Report: 1986* (Washington, DC: U.S. Government Printing Office, May 1987); and U.S. Department of Agriculture, Cooperative Extension Service, *Annual 4-H Youth Development Enrollment Report, 1987 Fiscal Year* (Washington, DC: U.S. Government Printing Office, 1988).

RETHINKING THE FEDERAL ROLE IN MATHEMATICS AND SCIENCE EDUCATION

In thinking of any policy problem, it is useful to identify what goals a system is intended to meet, what alternative actions need to be weighed, and how those actions can be implemented to fulfill system goals. In the case of elementary and secondary mathematics and science education, it is the last of these that is the most difficult. What needs to be done is much more obvious than determining *how* it is to be done. The Federal Government is historically at least one step removed from those who have the most direct influence on teaching and learning—families, teachers, schools, and students. The Federal Government can and does provide incentives and support for some actions rather than others; rethinking the mechanisms of Federal support affects the larger issue of the division of roles nationally in education.

Mathematics and science education are part of a much larger set of issues with national dimensions; all, however, are built from the ground up: neighborhood schools, locally elected school boards, and State governments. The tension between national and local priorities has a long history, but is ultimately an essential part of the American system of participatory democracy. Reconciling national and local visions should be regarded as the job of educational policymaking, not an obstacle in its way.

Today, a new phase of Federal interest in education is developing. It is based on the need to train a better quality work force as well as the need to ensure equity of educational provision to all young Americans. The heightened importance of these needs will require change in several areas, including organizational arrangements in schools and school districts, the upgrading of the teacher work force, and, ultimately, new spending. The real cost of elementary and secondary education is already rising. More important, the need to improve mathematics and science course offerings, introduce more experimental work in classrooms, extend informal learning opportunities, and fuel enrichment programs both in and out of school cannot be met by improvements in the existing system alone. In particular, greater use will be made of both individual and collective learning styles in class and out.

The special challenge to formal mathematics and science education is its ability to command adequate, but not excessive, attention relative to the vast number of other issues that arise in elementary and secondary education.

Appendixes

Appendix A

Major Nationally Representative Databases on K-12 Mathematics and Science Education and Students

Longitudinal[1]

1972-86 (continuing)[2]

National Longitudinal Study (National Center for Education Statistics, Department of Education). The cohort studied is 23,451 high school seniors (1972), enrolled in a total of 1,318 high schools. Followups were conducted in 1973-74, 1974-75, 1976-77, and 1979-80. A fifth followup was conducted on a subsample in 1986. In each followup, data were collected on high school experiences, background, opinions and attitudes, and life plans. Participants were subjected to a battery of achievement tests only in the first survey.

1980 - Present

High School and Beyond Survey (Center for Statistics, Office of Educational Research and Improvement, Department of Education). This survey contains two cohorts, starting in 1980. The first is a sample of 30,000 high school sophomores, and the second a sample of 28,000 high school seniors. These cohorts were enrolled in 1,015 public and private schools and are sampled every 2 years. The purpose of the survey is ". . . to observe the educational and occupational plans and activities of young people as they pass through the . . . system and take on their adult roles." The 1980 and 1982 surveys consist of a questionnaire (on attitudes, educational plans, family background, socioeconomic status data, and activities outside of school) and cognitive tests. The tests were developed by the Educational Testing Service and were intended to measure cognitive growth in three domains: verbal, mathematics, and science. The most recent followup data were released in 1988. No achievement tests were administered in the 1986 surveys.

Cross-Sectional

Various Years

National Assessment of Educational Progress (NAEP) (formerly administered by the Education Commission of the States and now by Educational Testing Service, Inc. (ETS)). NAEP is a congressionally mandated program, funded by the Office of Educational Research and Improvement of the Department of Education, that assesses national achievement in education, including science and mathematics.[3] ETS now refers to it as "The Nation's Report Card." The most recent science and mathematics assessments were in 1985-86 (published in June 1988); previous assessments were in 1981-82, 1977-78, 1972-73, and, for science only, 1969-70. Each science test measured both science attitudes and achievement. The 1981-82 and 1985-86 assessments addressed attitudes to science in more detail than before and the 1985-86 assessment, for the first time, asked for background information on science experiences out of school and on what is being taught in science classrooms. In 1985-86, teachers were also asked to provide information on their training and experience, instructional methods, and their intended curriculum. To conduct the assessment, NAEP identifies a stratified probability sample of schools and tests students in three age groups: 9, 13, and 17. Anywhere between 60,000 and 90,000 students take NAEP tests, although any given test item is taken only by 2,600 individuals. (ETS terms the technique "Balanced Incomplete Block Spiralling.") After assuming responsibility for NAEP, ETS introduced a degree of cohort matching into its assessment design.[4] Co-

[1]Dates refer to time of data collection. Post-1986 databases currently under construction and subject solely to primary analysis are excluded. For titles of ongoing database projects supported by the National Science Foundation, see *Summary of Active Awards, Studies and Analysis Program* (Washington, DC: March 1988).

[2]General note on the two surveys in this category: although they are intended to be similar and comparable, they differ. High School and Beyond includes data from parents and teachers, as well as from high school transcripts (allowing relationships between course-takings, achievement, and destinations to be estimated). The definition of "Hispanic" has also changed between the two studies, affecting comparisons of minority performance. The low response rate to the National Longitudinal Study limits its usefulness. The 1988 National Educational Longitudinal Study will reduce such limitations.

[3]The National Assessment of Educational Progress was initiated in 1969. Note that the 1981-82 science assessment was not funded by the Department of Education or conducted by the Educational Testing Service, but through a special grant from the National Science Foundation to the University of Minnesota.

[4]The purpose of cohort matching is to improve the confidence that can be placed in conclusions about changes in student achievement through time. The matching will work by applying the same sampling technique to 9-year-

(continued on next page)

hort matching has not been applied to assessments before 1985, which limits the ability of NAEP to support conclusions about improvements or declines in science and mathematics achievement over time.

1977 and 1985

National Surveys of Science and Mathematics Education, sponsored by the National Science Foundation and conducted by Iris Weiss of Research Triangle Institute. The survey requested science and mathematics course offerings and enrollment (by race, ethnicity, and sex), science facilities and equipment, instructional techniques used, and teacher training. All data were self-reported. The 1985 survey covered 425 schools, public and private, including 6,000 teachers, and was stratified by grades: K-3, 3-6, 7-9, and 10-12. The survey was published in November 1987. The data do not address achievement.

1985

Survey of High School Teachers and Course Offerings, conducted by the National Science Teachers Association (NSTA). The survey was based on an "Official U.S. Registry of Teachers" maintained by NSTA, which is intended to list all science teachers of grades 7-12. All private and public secondary school principals were asked to name all their science teachers and the number of classes of each type of science that the teachers will teach during the coming school year. A stratified random sample of 2,211 high schools was culled from 26,000 responses, derived in turn from a total pool of 48,427 forms mailed. Stratification was by seven ranges of school size, three grade structures (K-12, 7-12, and either 9-12 or 10-12), and by public or private status. Data have been used by NSTA to estimate the number of sections of each type of science course taught, the number of high schools that offer either no physics, chemistry, or biology courses, the number of teachers teaching "in-field" and "out-of-field," and the teaching load of average physics, biology, and chemistry teachers.

Annually

The American Freshman, conducted by the Cooperative Institutional Research Program (CIRP), University of California at Los Angeles (UCLA). CIRP and UCLA's Higher Education Research Institute annually survey incoming freshmen in full-time study in colleges and universities. Some longitudinal followup studies are attempted to track students 2 and 4 years after college entry. All incoming freshmen are surveyed and the data are stratified by type of college, public or private control, and the "selectivity level" of the institution. The survey instrument solicits data on high school background, including Scholastic Aptitude Test or American College Testing program score and grade point average, intended major and educational destination, career intentions, financial arrangements, and attitudes.

Cross-Sectional and International[5]

1980-82

Second International Mathematics Study (SIMS) (U.S. Co-ordinator: College of Education, University of Illinois at Urbana-Champaign. Published as *The Underachieving Curriculum*, January 1987). Data were collected in 1980-82 and covered 20 countries. The U.S. survey covered 500 classrooms in 2 populations: grades eight and those completing secondary school and taking a large number of mathematics courses (defined as 12th graders taking college preparatory mathematics). Both SIMS and the Second International Science Study (SISS) focused on three stages of the educational process: the intended curriculum (what is in the textbooks), the implemented curriculum (what the teacher actually does in class and how), and the attained curriculum (what the students learned). The intended curriculum is based on a content analysis of textbooks, while the implemented curriculum was measured by asking teachers to complete questionnaires. The attained curriculum was tested by multiple-choice achievement tests. SIMS attempted to study the relationship between what happens in schools and what students learn, so the achievement batteries were administered twice: once at the beginning and once at the end of each school year.

1983 and 1986

Second International Science Study (U.S. Coordinator: Teachers College, Columbia University). SISS, similarly to SIMS, attempted to address what it termed the "intended," "translated," and "achieved" curricula for science instruction in 25 countries. SISS studied

(continued from previous page)

olds in 1985-86 as will be applied to 13-year-olds in 1989-90 and 17-year-olds in 1993-94. The individuals tested will not be identical, however.

[5]The following two studies were conducted under the aegis of the International Association for the Evaluation of Educational Achievement, a nongovernmental association of educational researchers. The first international science study was in 1970 and the first international mathematics study was in 1964. Both the second mathematics and science studies are "deeply" stratified probability samples of different types of schools, made under internationally agreed on and applied guidelines. The processing and interpretation of results have been hampered by lack of funding and of a central location for performing the necessary work.

both public and private schools, and surveyed a total of 2,000 U.S. students, broken into four populations: grade 5; grade 9; grade 12 taking second-year physics, chemistry, and biology; and grade 12 taking no science (1983 only[6]). In the United States, 125 schools were surveyed and the tests addressed science achievement, attitudes, "a science learning inventory," and a word knowledge test. Background data on the school and about teachers were also requested. Some items common with the First International Science Study were included; at grade 5, there were 26 such items, and at grade 9, there were 33. Both grades five and nine showed an improvement over performance in 1970. Results from the two grade 12 populations in the United States have not been published. Data from other countries are still being processed in Stockholm, funded by the Swedish Ministry of Education and the Bank of Sweden. The 1986 survey addressed process skills and content, as well as teacher activities and other factors in the school environment that might affect achievement. Data were collected from students, science teachers, and school administrators. Process and achievement skills were tested for both grades five and nine. Achievement was be tested in these groups: 10th grade biology, 11th grade chemistry, and 12th grade students with more than 1 year of study in one or more of physics, chemistry, or biology. Technical aspects of this survey, including data collection but not instrument design, were contracted to the Research Triangle Institute.

[6]Data for 1983 are suspect due to only a 30 percent response rate.

Appendix B
Mathematics and Science Education in Japan, Great Britain, and the Soviet Union

During recent years, a steady stream of international comparisons of elementary and secondary education has painted an increasingly bleak picture of the deficiencies of American mathematics and science education. Depressing comparisons with teaching practices in Japan, Taiwan, Hong Kong, and South Korea have been seized on by many of those pressing for the United States to address its crumbling competitiveness. The source of competitive advantage is often referred to as "human resources" or "human capital"—skilled, talented, and flexible workers. For example, a recent commentary noted that international comparisons of mathematics education " . . . typically depict Korean 10-year-olds working out the Four-Color-Map Problem in their heads while Americans of the same age struggle to do double-digit multiplication without removing their socks."[1]

Problems in Making International Comparisons

While international comparisons do point to significant differences in mathematics and science course offerings, curricula, and teaching, it is important to bear in mind two major problems in any sort of policy-oriented comparisons among cultures and countries:

- developing a sufficiently accurate explanation of the *causes* of observed differences, which are very often rooted in what are, to the outsider, opaque cultural and social differences in the roles of families, business, the State, law, and education; and hence
- determining what aspects of other systems could readily be *appropriated and transferred* across cultures and societies, and which would be foolish or even counterproductive to consider transferring.

For example, the United States could adopt a national curriculum, but such a move would be resisted strongly by many policymakers. Such a move would threaten the fragile compacts between national and local autonomy stipulated in the Federal as well as State constitutions. Further, a "national curriculum" would likely consist of little more than a lowest common denominator of topics defined by special interest groups and argued out line by line in highly partisan congressional debates. Rather than providing models to be emulated, the ultimate value of doing international comparisons may be to provide a kind of "mirror" in which to examine and better understand the reasons for well-entrenched, culturally rooted American educational practices and policies.

On an analytical level, it is difficult to make sound international comparisons in education unless studies are designed to compare "like with like" and to collect enough data to build a picture of overall educational capacity—teachers, students, and schools—in each country. In considering the high school students' exposure to mathematics and science, for example, it is important to note that the American school system is designed to retain all students to age 18 (and actually succeeds in enrolling about three-quarters of this group), whereas schools in other countries typically enroll a much more select group of students in the 14- to 18-year-old range.

These caveats aside, it is generally agreed that the American education system devotes relatively less time to mathematics and science education compared with other countries; estimates are that American students spend only one-third to one-half as much time on learning science as their peers in Japan, China, the U.S.S.R., the Federal Republic of Germany, and the German Democratic Republic.[2] Significant differences in the mathematical progress of children in selected cities in the United States, Taiwan, and the People's Republic of China have been found from the elementary grades. Japanese kindergarten children already surpass American children in their understanding of mathematical concepts.[3] It is evident that differences are across the entire educational system of each nation.

Japan

The country with which commentators most enjoy making international comparisons of mathematics and science education is Japan.[4] Japanese children study

[1] Edward B. Fiske, "Behind Americans' Problems With Math, A Question of Social Attitudes," *New York Times*, June 15, 1988, p. B8.

[2] F. James Rutherford et al., *Science Education in Global Perspective: Lessons From Five Countries* (Washington, DC: American Association for the Advancement of Science, 1985).

[3] Harold W. Stevenson et al., "Mathematics Achievement of Chinese, Japanese, and American Children," *Science*, vol. 231, Feb. 14, 1986, pp. 693-699.

[4] The following is based largely on William K. Cummings, "Japan's Science and Engineering Pipeline: Structure, Policies, and Trends," OTA con-

far more mathematics and science than American children, and more of them emerge from schools with a greater degree of scientific and technological literacy than do American children. Japanese institutions of higher education can draw from an exceptionally well-qualified crop of students. In addition, many go on to science and engineering majors; it is estimated that, in proportion to its population, Japan produces as many scientists and twice as many engineers as does the United States.

Japanese education resides within the cultural milieu of Japan. A well-known recent study explored the importance of families in mathematics education. While Japanese families considered poor performance in mathematics to be a consequence of lack of effort, American families more often attributed success to innate ability, despite poor teaching.[5]

Japanese students typically attend school for 240 days per year (compared with 190 in the United States), because they work a half day on Saturdays. It is believed that they spend a greater proportion of class time on academic activities.[6]

In elementary schools, the science curriculum includes matter and energy, living things and their environments, and the Earth and the universe. These themes are often reinforced by educational television programs broadcast by the Japan Broadcasting Association and coordinated with the curriculum.[7] Students often go on field trips with their school. It is widely reported that Japanese mathematics and science curricula demand more from their students than those in the United States: Japanese students simply cover more ground. Although Japanese and American elementary school students spend a similar amount of time on mathematics and science, from six to eight periods per week, the time is far more intensively used in Japanese schools.

tractor report, October 1987. See also U.S. Study of Education in Japan, *Japanese Education Today*, OR 87-500 (Washington, DC: U.S. Department of Education, January 1987); John Walsh, "U.S.-Japan Study Aim Is Education Reform," *Science*, vol. 235, Jan. 16, 1987, pp. 274-275; Debra Viadero, "Japan and U.S. Release Assessments of Each Other's Education Systems," *Education Week*, Jan. 14, 1987, pp. 9, 19; Willard J. Jacobson et al., *Analyses and Comparisons of Science Curricula in Japan and the United States*, Second IEA Science Study (New York, NY: Columbia University Teacher's College, 1986); and Wayne Riddle, Congressional Research Service. "Public Secondary Education Systems in England, France, Japan, the Soviet Union, the United States, and West Germany: A Comparative Analysis," EPW 84-770, Issue Brief, 1984, pp. 16-21.

[5]Stevenson et al., op. cit., footnote 3, p. 697; U.S. Study of Education in Japan, op. cit., footnote 4, p. 3.

[6]The amount of time spent on a subject is one, somewhat crude, indicator of the amount of learning in that subject. But the amount of learning also depends on the efficiency of that learning, which depends in part on the quality of teaching methods employed and, in the elementary years, the links with other subjects. Learning in mathematics and science by elementary school children depends in part on their reading and writing abilities.

[7]Tomoyuki Nogami et al., "Science Education in Japan—A Comparison With the U.S.," mimeo prepared for the National Science Teachers Association 35th National Convention, Mar. 27, 1987, pp. 2-3.

Research data suggest that there is less variation among students in mathematics and science learning than in the United States. In part that is because Japanese elementary schools are not grouped or tracked by ability, and the use of mixed-ability cooperative learning groups, or *han*, is very common. But it is also due to the assumption that everyone can and must be competent in these subjects. A recent book notes that:

> It is simply taken for granted . . . that every child must attain at the very minimum "functional mathematics," that is, the ability to perform mathematical calculations in order to accomplish requirements successfully at home or work.[8]

Japanese lower secondary schools, which cover grades seven to nine, are similar to American junior high schools, but have few or no electives. Students are required to wear uniforms and adopt a more serious and disciplined approach to work than they had in elementary schools. There is still no sorting by ability, although the use of cooperative learning groups is rare. Upon completion of lower secondary schools, at the end of ninth grade, all students have taken some elementary geometry, trigonometry, algebra, and probability and statistics. They have devoted between 6 to 8 out of 30 weekly class periods to mathematics and science. In science, students take a variety of general science topics, including biology, chemistry, physics, and earth science. Teachers specialize in a subject area and teach only that area. The pressure to get into a good upper secondary school leads students in lower secondary schools to be highly competitive and neurotic, and the consequent pressure to succeed is often cited as a cause of teenage suicide. Many students attend out-of-school *juku*, which are coaching lessons that prepare them for examinations for entry to upper secondary schools, mostly concentrating on English and mathematics.

The final 3 years of school, upper secondary school, are quite different from the preceding 9. Entry to this level is on the basis of lower secondary school records and common entrance examinations administered by the local prefecture; schools can be highly selective, and there is considerable variation in the courses that students take. Attendance at upper secondary schools is voluntary, and tuition is charged. About 90 percent of young people attend them. Typically, several upper secondary schools serve students within a given neighborhood and there is a clear ranking of prestige among them. Some upper secondary schools specialize in academic-preparatory programs and other vocational programs; only about 30 percent offer both programs.[9] While the post-war reforms at first embraced

[8]Benjamin Duke, *The Japanese School: Lessons for Industrial America* (Westport, CT: Praeger Special Studies, 1986), p. 82.

[9]U.S. Study of Education in Japan, op. cit., footnote 4, p. 41.

the neighborhood comprehensive high school principle, parents often lobbied officials to send their children to the better neighborhood schools and upper secondary schools now are definitely not equal.

Within Japanese upper secondary schools, students select a given course of classes, and it is very difficult to change course or take classes outside those specified for that course. There are few or no electives. Typically, academic-general and vocational-specialized courses are offered, although both often have common coursework in the first year. Within the academic-general course there is often a branch at the end of the first year of upper secondary school, at which point those planning science and engineering majors are separated from those planning arts majors. Students are often further sorted by ability levels within each course. About two-thirds of the students take academic-general courses, one-quarter take vocational courses, and the remainder enter specialized colleges of technology or training schools, or enter the work force directly.

Those planning science and engineering majors take a total of 102 credits over 3 years, of which 18 are in mathematics and 16 are in science; these credits include calculus, physics, chemistry, and biology. In grade 10, students spend 10 out of 34 hours per week in mathematics and science courses, rising to 14 and 18 hours in grades 11 and 12, respectively. In grade 12, the science-bound take 5 hours per week of integral and differential calculus as well as both physics and chemistry. In total, one-half of those in academic high schools are in science courses. Nevertheless, the core courses required for both the arts and the vocational-specialized courses are sufficiently demanding in mathematics and science that some students who have taken these courses can still compete for entry to college science and engineering programs. The uniformity of classes means that Japanese students have no equivalent of the advanced placement examinations, and must all start with freshman mathematics science programs at college. The upper secondary school curriculum is very demanding, and some students fall behind and lose interest. The net effect is that, while the proportion of 22-year-olds that receive baccalaureates in the United States and Japan is about the same (23 to 24 percent), about one-quarter of these in Japan are in natural science and engineering as compared to 15 percent in the United States.

Japanese public upper secondary schools are complemented by a number of private upper secondary schools, in part because only a limited number of public schools were built after the war and the education in lower grades was emphasized. About one-quarter of upper secondary schools are private. While some of these schools are highly selective, others enroll those who failed to qualify for the limited number of places in the public sector. Public schools generally carry more prestige.

For the college-bound, the upper secondary years are extremely demanding as students prepare to take college entrance tests. In the Japanese system, there is well-defined ranking of higher education institutions and a student's decision to enroll in a particular institution will have, through contacts with students and professors, a great effect on his or her later career, job prospects, and life. The college education that students receive is relatively unchallenging. The great competition to enroll in the "right" university has created pressure for a very intensive academic curriculum and the extensive use of examinations to sort and prepare students for college entrance in the upper secondary school years. Students often take both the nationally administered First Stage Standardized Achievement Examination and examinations set by the particular university to which they are applying. These examinations test only factual recall and include no testing of skills with experiments and the process of construction of new scientific knowledge. In preparing for the examinations, students often enroll in *yobiko* which are similar to the *juku* at lower secondary level.

In many ways, it is ironic that the Japanese school system does so well in mathematics and science whereas the American system has problems, because Japan's schools were reorganized along American lines. The ultimate aim was to democratize and demilitarize Japan during the post-war American occupation. Japan's schools have also made good use of curricula and instructional material developed in America. Japanese schools are run by about 50 prefectural-level school boards and 500 local school boards which, unlike many of their American counterparts, are filled by people appointed from above rather than elected. The national Ministry of Education, Science, and Culture (monbusho) prescribes curricula, approves textbooks, provides guidance and funding, and regulates private schools. The prefectural boards of education appoint a prefectural superintendent of education, operate those schools which are established by the prefectures (primarily upper secondary schools), license and appoint teachers, and provide guidance and funding to municipalities. Municipal boards operate municipal schools, choose which textbooks to use from those approved by monbusho, and make recommendations about the appointment and dismissal of teachers to the prefectural board.

The cost of education is shared by the various tiers of government and, at later stages, by parents directly. The Japanese government pays about half of the cost, including some subsidies to private schools. Special budget equalizing regulations direct the central government to augment the spending of poorer districts up to a minimum level; there still remains a 60 percent variation in average per pupil expenditure among school districts. There is a uniform national pay scale for teachers, with a modest starting salary and steady annual increments that can triple a teacher's income after 20 years of service. Overall salaries are competitive with other occupations, and teaching is an attractive job; there are normally at least enough applicants for teaching positions, even in mathematics and sciences. Teaching remains one of the limited number of professional occupations that are readily available to women, but most teach only in elementary schools; only 18 percent of teachers in upper secondary schools are female. Most teachers have degrees in single academic disciplines other than education, although substantial numbers have no degree at all. Each prefecture and large city has an education center for its teachers, which provides inservice training and conducts educational research.

Curricula, in principle, are controlled by local school boards, but the central Ministry of Education sets standards in a "National Course of Study" and approves textbooks. In practice, it is believed that there is considerable uniformity of curricula. Curricula emphasize mastery of key subjects, such as mathematics and science, and make few concessions to individual learning styles, predilections, and idiosyncracies. It is a belief in Japanese culture that creativity is only possible once fundamentals are mastered.

Just as Americans admire the uniformity and excellence of the Japanese system, many Japanese are searching for ways to make their system less monolithic and competitive, and more respectful of individual creativity, particularly as the country is now stressing the development of a basic research capability.[10] Japanese students are schooled to master factual material rather than analysis, investigation, or critical thinking. Teacher lectures from textbooks are very common, although considerable use is also made of laboratory work in science.[11] Similar programs of educational reform analysis and activity exist in Japan and the United States; the United States program is often taken as a model.[12] The Prime Minister's National Council on Education Reform called in September 1987 for the school system to put more emphasis on diversity and creativity, but the council has been split by internal controversy over the shape of possible reforms. It is believed that these calls may lead some prefectures to combine lower and upper secondary schools and to put a reduced emphasis on examination-based sorting.[13]

Great Britain

Most observers of Great Britain's system of mathematics and science education have concluded that it is very good for those who plan college study in science and engineering, but comparatively weak for the rest.[14] Students, by international standards, specialize at a very early age, and are offered few opportunities to shift interests. Preparation for examinations, which are externally set and assessed, is the dominant activity for those college-bound. Enrollment in higher education institutions is restricted both by financial mechanisms and fairly strict entry requirements, and the proportion of British 18- to 22-year-olds that enroll in college may be the smallest in any developed nation.

Like the United States, comparatively little science is taught in British elementary schools and there are few teachers qualified or motivated in the subjects. The emphasis instead is on mathematics. Most students begin science classes in one or more of the fields of physics, chemistry, and biology in the sixth grade, and continue them to the end of eighth grade. In grades 9 and 10, students choose a limited menu of courses to specialize in for the purposes of taking nationally administered school-leaving examinations (now known as the General Certificate of Secondary Education) at the end of 10th grade. Those college-bound in science and engineering might take six to eight subject exams, of which three or four are in mathematics and science subjects. Enrollment in grades 11 and 12 is voluntary, and is normally restricted to those preparing for nationally administered college entrance examinations (known as Advanced-level, or A-level) examinations. At this stage, students normally specialize in only two or three subjects, which tend to be related to each other, although proposals have often been made to expand this number in the interests of giving college students a "broader education." In these grades, the science- and engineering-bound can take only mathematics and science subjects, if they choose.

Entry to colleges and universities in Great Britain is granted on the basis of grades and the number of passes awarded in the A-level examinations, and most

[10] The Japanese parallel to the U.S. Study of Education report is Japanese Study Group, Japan-United States Cooperative Study on Education, "Educational Reforms in Japan," mimeo, January 1987, which studied aspects of U.S. educational reform that might translate to Japan, especially the transition between secondary and higher education.

[11] U.S. Study of Education in Japan, op. cit., footnote 4, p. 34.

[12] Ibid., pp. 63-67; and Walsh, op. cit., footnote 4.

[13] David Swinbanks, "Reform Urged for 'Hellish' Japanese Education System," *Nature*, vol. 326, Apr. 16, 1987, p. 634.

[14] See Joan Brown et al., *Science in Schools* (Philadelphia, PA: Open University Press, 1986).

institutions make offers to students conditional upon their achievement of certain grades. A minimum of two passes in two subjects at A-level is normally required, and entitles the student to free tuition and access to a grant program for living expenses while at the university.

The Soviet Union

In many ways, the mathematics and science education system in the Soviet Union is similar to the American and British systems.[15] Each is a very uneven system, geared primarily to training those who will become professional scientists and engineers. The bulk of the school population learns relatively little and is left largely alienated from science and engineering. The extremes in the Soviet Union, however, are quite distinct. The layer of top quality students, which ranks internationally with the best, is very thin; there is a steep drop-off below that level.

The education system in the Soviet Union, like that in the United States, Great Britain, and Japan, is undergoing reforms. The impetus for these reforms is the new national desire to improve the efficiency of industry.

Soviet elementary and secondary education is split into three phases:

- primary or elementary education, grades 1-3
- "incomplete" secondary education, grades 4-8; and
- "complete" secondary education, grades 9 and 10.

Most students complete eighth grade, normally achieved at age 15, and then choose among several different paths. Some continue in general education secondary schools and are most likely to enter higher education, while others enter the work force but continue their secondary education via evening or correspondence courses. Higher education is open to students from each of these routes, but most college students come from those who enrolled full time in general education secondary schools. A third path after eighth grade is to enroll in a technical or vocational school. These schools are primarily intended to train future skilled workers and technicians. For the most talented students, there are a number of specialized schools, often residential, some of which specialize in mathematics and science.

Entry to higher education in the Soviet Union is competitive, with an approximate oversupply of secondary school graduates of between two and three times the number of higher education places available. Admission to higher education institutions is primarily by means of examinations. Political considerations are often important, and students who participate in extracurricular social service activity, such as agricultural work, are often given some preference in admissions and award of stipends. Although only about 20 percent of students go on to higher education, about half of these are in science and engineering, with the effect that a larger proportion of each birth cohort graduates in sciences and engineering than in the United States.

Reforms in progress include a requirement for students to begin full-time education a year earlier, at age 6 rather than age 7. The extra year will allow time for study of additional subjects such as labor education and computer science. At the moment, computers are very rarely used in classrooms, but their use has become a national priority. All students will have required courses in information science, but it is likely that many of these classes will be taught without hardware. To improve the skills of the work force, all students in general secondary education learn manual labor skills, even those who go on to higher education. The current phase of curriculum reform is designed to introduce less demanding, but more comprehensive, curricula in many subjects, including mathematics and science. This last step is ironic, because the intensity of the traditional mathematics and science curriculum is often praised by American commentators. More money is being allocated to education; teachers' salaries are being increased and merit pay arrangements are being introduced.

Under the reforms, closer links are being forged with both higher education institutions and local firms and enterprises. All schools are being paired with neighborhood factories and businesses: enterprises will be required to provide financial and material assistance to schools and hope to receive better trained workers. Many higher education institutions are signing agreements with secondary schools, under which college faculty will assist in secondary school teaching. Students will be given access to research facilities at colleges, and the colleges will admit a specified number of students to higher education. Research laboratories are also being encouraged to participate in such agreements. So far, 30 of about 900 higher education institutions have signed agreements of this kind. While reforms of this kind should improve the quality of education of future scientists and engineers, they will force students to specialize at a younger age, often 15 or 16, than has been the case.

As a further part of the Soviet reform program, more students will be encouraged to enter technically oriented vocational programs in grades 9 and 10 rather

[15]The following is based on Harley Balzer, "Soviet Science and Engineering Education and Work Force Policies: Recent Trends," OTA contractor report, September 1987.

than general education secondary programs. Although the training these students receive is supposed to be equivalent to that in general education secondary programs, this step will probably reduce the already oversubscribed pool of qualified high school graduates planning to enter higher education. College entry standards have also been tightened during the last 8 years, except in fields such as computer science, and biology, which have been targeted for expansion.

Conclusions

While mathematics and science education in other countries contains many features to be admired, including the commitment to these subjects, the emphasis on academic learning, and the geographical equalization of learning opportunities, each system contains some features that disturb many American educators. They include unswerving allegiance to the mastery of facts rather than creativity and individual expression and the extensive use of lectures. Other unappealing features are the limited number of women teaching in upper secondary schools, the central control over many aspects of teaching, and the teaching of science and technology literacy to a select few rather than the full cohort of students. Above all, many other countries have a greater degree of centralization in educational decisionmaking than exists in the United States. Schools and school districts need to identify practices that are transferable, but also need to be extremely cautious of the cultural and social assumptions underlying some of these practices.

Appendix C

OTA Survey of the National Association for Research in Science Teaching

The National Association for Research in Science Teaching (NARST) is an organization of university researchers in mathematics and science education. In June 1987, OTA mailed questionnaires to the American membership of NARST, asking for their opinions on a range of current and past Federal programs designed to improve precollege mathematics and science education. Of 500 questionnaires mailed out in this informal, one-shot survey, 135 were returned for a response rate of just over 30 percent.

The survey was designed to give OTA a general impression of the opinions held by individuals knowledgeable about previous Federal efforts in this area. NARST includes many active participants and evaluators of previous Federal mathematics and science education programs; their responses provided valuable, often first-hand, accounts of programs such as National Science Foundation (NSF) summer institutes and curriculum development programs. Respondents also had extensive familiarity with other local, State, and NSF-sponsored programs that are listed below. OTA recognizes that the members of NARST are neither representative of the broad population of researchers and teacher educators in mathematics and science education, nor are the responses statistically representative of the entire NARST membership.

OTA solicited opinions on the effectiveness of the following programs:

- grants for equipment and supplies under the National Defense Education Act (NDEA) of 1958;
- NSF summer institutes and other inservice training for teachers;
- NSF summer institutes for students and other research participation programs for students;
- NSF-funded new curriculum programs;
- assistance for magnet schools for racial desegregation purposes;
- funds allocated through Title II of the Education for Economic Security Act of 1984; and
- support for informal science education, including educational television and science and technology centers.

Of these programs, NSF teacher institutes, research participation for students, and curriculum development programs received the highest ratings. Many respondents thought that significant lessons had been learned from the teacher institutes and curriculum development programs, such as the need to ensure that participating teachers are given followup training and that training covers both subject knowledge and ideas for teaching mathematics and science in real-life situations. Magnet school programs were the least highly rated, although less than one-half of the respondents cited these programs at all.

OTA also asked respondents to identify other Federal programs that they thought had had a positive effect on mathematics and science education. These nominations provide a fairly comprehensive overview of the kinds of Federal programs that have been attempted since passage of the NDEA:

- NSF-funded Institutes for Science and Mathematics Supervisors;
- funds for Research on Teaching and Learning of Science and Mathematics;
- NSF-funded Chautauqua Institutes for Teachers;
- the Department of Education's National Diffusion Network for the dissemination of effective curricula materials;
- Programs for Metric Education;
- the Department of Education's Fund for the Improvement of Post-Secondary Education;[1]
- Presidential Awards for Teaching Excellence in Mathematics and Science;
- the National Assessment of Educational Progress, funded by the Department of Education;
- NSF's Project SERAPHIM;
- the National Aeronautics and Space Administration's program for sending astronauts and scientists to schools, and a similar program funded by NSF;
- the Clearinghouse for Research in Science, Mathematics, and Environmental Education that is part of the ERIC system;
- National Sea Grant College Marine Education Activities;
- the Department of Energy's Honors Science Program; and
- NSF's program of Resource Centers for Science and Engineering.

[1]Despite its name, this source has funded some precollege programs, such as the *Family Math* series of books and materials from the EQUALS program at the Lawrence Hall of Science, University of California at Berkeley.

Several respondents made noteworthy observations regarding:

- the importance of educating guidance counselors, principals, and school board members about the problems of science education, particularly to convey that science is a collection of facts as well as a way of exploring and examining the world, and that the interest of many students in science needs to be developed;
- the diminutive size of Federal mathematics and science education programs in relation to the size of the problems;
- the fact that past programs have not done a good job at distinguishing between programs for training future scientists and engineers and programs for boosting technological literacy. NSF programs have often been aimed at the former, and have made too great a use of working scientists who have limited appreciation of the culture of schools and the need to involve parents, school boards, and administrators if improvement is to be lasting and meaningful;
- the need for ongoing training and support for programs, once mounted. In many cases, programs have attempted too much too quickly, and have failed to follow through, allowing the status quo to be reasserted;
- the increased use of science specialists and consultants to be shared among schools or even school districts;
- the need for more emphasis on elementary mathematics and science education, where the battle is won or lost, rather than on secondary education when too many deficiencies are already irrevocable; and
- the recognition that good teachers are an absolute precondition to improvements in other areas, such as curriculum, equipment, and testing. Science and mathematics teacher education programs at universities need to be updated and fortified.

Appendix D

Contractor Reports

Full copies of contractor reports prepared for this project are available through the National Technical Information Service (NTIS), either by mail (U.S. Department of Commerce, National Technical Information Service, Springfield, VA 22161) or by calling them directly at (703) 487-4650.

Elementary and Secondary Education (NTIS order number *PB 88-177 944/AS*)

1. "Images of Science: Factors Affecting the Choice of Science as a Career," Robert E. Fullilove, III, University of California at Berkeley
2. "Identifying Potential Scientists and Engineers: An Analysis of the High School-College Transition," Valerie E. Lee, University of Michigan

International Comparisons (NTIS order number *PB 88-177 969/AS*)

1. "Japan's Science and Engineering Pipeline: Structure, Policies, and Trends," William K. Cummings, Harvard University
2. "Soviet Science and Engineering Education and Work Force Policies: Recent Trends," Harley Balzer, Georgetown University